The Olympic Games and the Environment

Global Culture and Sport

Series Editors: **Stephen Wagg** and **David Andrews**

Titles include:

Mahfoud Amara
SPORT, POLITICS AND SOCIETY IN THE ARAB WORLD

Aaron Beacom
INTERNATIONAL DIPLOMACY AND THE OLYMPIC MOVEMENT
The New Mediators

Mike Dennis and Jonathan Grix (*editors*)
SPORT UNDER COMMUNISM
Behind the East German 'Miracle'

John Harris
RUGBY UNION AND GLOBALIZATION
An Odd-Shaped World

Graeme Hayes and John Karamichas (*editors*)
THE OLYMPICS, MEGA-EVENTS AND CIVIL SOCIETIES
Globalization, Environment, Resistance

John Karamichas
THE OLYMPIC GAMES AND THE ENVIRONMENT

Jonathan Long and Karl Spracklen (*editors*)
SPORT AND CHALLENGES TO RACISM

Roger Levermore and Aaron Beacom (*editors*)
SPORT AND INTERNATIONAL DEVELOPMENT

Pirkko Markula (*editor*)
OLYMPIC WOMEN AND THE MEDIA
International Perspectives

Peter Millward
THE GLOBAL FOOTBALL LEAGUE
Transnational Networks, Social Movements and Sport in the New Media Age

---

Global Culture and Sport
Series Standing Order ISBN 978–0–230–57818–0 hardback
978–0–230–57819–7 paperback
(*outside North America only*)

You can receive future titles in this series as they are published by placing a standing order. Please contact your bookseller or, in case of difficulty, write to us at the address below with your name and address, the title of the series and the ISBN quoted above.

Customer Services Department, Macmillan Distribution Ltd, Houndmills, Basingstoke, Hampshire RG21 6XS, England

---

The Olympic Games and the Environment

John Karamichas
Queen's University, Belfast, UK

First published 2013 by
PALGRAVE MACMILLAN

Palgrave Macmillan in the UK is an imprint of Macmillan Publishers Limited, registered in England, company number 785998, of Houndmills, Basingstoke, Hampshire RG21 6XS.

Palgrave Macmillan in the US is a division of St Martin's Press LLC, 175 Fifth Avenue, New York, NY 10010.

Palgrave Macmillan is the global academic imprint of the above companies and has companies and representatives throughout the world.

Palgrave® and Macmillan® are registered trademarks in the United States, the United Kingdom, Europe and other countries.

ISBN 978–0–230–22861–0

This book is printed on paper suitable for recycling and made from fully managed and sustained forest sources. Logging, pulping and manufacturing processes are expected to conform to the environmental regulations of the country of origin.

A catalogue record for this book is available from the British Library.

A catalog record for this book is available from the Library of Congress.

Transferred to Digital Printing in 2013

For Angelica

Contents

List of Figures and Tables

Figures

Tables

Preface and Acknowledgements

In many respects, this book has been an Olympian challenge. I am not going to elaborate the issues surrounding this view. It suffices to say that I learnt a lot in this journey; and I owe a big thanks to many people who gave their time in discussing my ideas and helping me in various ways to complete this book. This book would have been near impossible without them. As I do not want to risk forgetting any of them, I will not provide an individual name list. I would, though, like to give special thanks to the School of Sociology, Social Policy and Social Work at Queen's University Belfast for allowing me to have a semester dedicated to the research and writing of this book. I also wish to thank Philippa Grand, Olivia Middleton and Andrew James at Palgrave Macmillan for their invaluable help in various stages. Last but not least, a heartfelt thanks to Amanda for bringing into this world and looking after Angelica, our daughter.

John Karamichas
Moira, Northern Ireland

List of Abbreviations

ALP	Australian Labour Party
ANU	Australian National University
ARF	Athens Regulatory Framework
ATHOC	Athens Organizing Committee
BMBS	Beijing Municipal Bureau of Statistics
BOBICO	Beijing Olympic Games Bidding Committee
BOCOG	Beijing Organizing Committee for the Olympic Games
BRICS	Brazil, Russia, India, China, South Africa
CCC	Committee on Climate Change
CCP	Chinese Communist Party
CCS	carbon capture and storage
COHRE	Centre On Housing Rights and Evictions
CPRS	Carbon Pollution Reduction Scheme
CSL	Commission for Sustainable London
Defra	Department for Environment, Food and Rural Affairs
DPRK	Democratic People's Republic of Korea
EIA	Environmental Impact Assessment
EIS	Environmental Impact Statement
EJM	Environmental Justice Movement
EM	ecological modernization
ENGO	environmental non-governmental organization
EPB	Environmental Protection Bureau
EPUK	Environmental Protection UK
ES	environmental sustainability
ETS	Emissions Trading Scheme
EU	European Union
FDI	foreign direct investment
FoE	Friends of the Earth
FYP	five-year plan
GHG	greenhouse gases
GONGO	government organized non-governmental organization
HEP	Human Exemptionalism Paradigm
HLS	Higher Level Stewardship
ICM	Independent Communications and Marketing

IMDs	indices of multiple deprivation
IOC	International Olympic Committee
IPCC	Intergovernmental Panel on Climate Change
IPPR	Institute of Public Policy Research
LAOOC	Los Angeles Organizing Olympic Committee
LAOS	Popular Orthodox Rally
LA21	Local Agenda 21
LDA	London Development Authority
LLP	Left-Libertarian Party
LOCOG	London Organizing Committee for the Olympic Games
LULU	locally unwanted/undesirable land use
MCC	Manual for Candidate Cities
MGAS	Manor Garden Allotments Society
ND	New Democracy
NDRC	National Development and Reform Commission
NEP	New Ecological Paradigm
NGO	Non-Governmental Organization
NHE	New Human Ecology
NLLDC	New Lammas Land Defence Committee
NMC	new-middle class
NNR	National Nature Reserve
NOC	National Olympic Committee
NSM	New Social Movement
NSW	New South Wales
OCA	Olympic Coordination Authority
OCOG	Organizing Committee for the Olympic Games
ODA	Olympic Delivery Authority
OECD	Organization for Economic Cooperation and Development
OGI	Olympic Games Impact
PASOK	Panhellenic Socialist Movement
PRC	People's Republic of China
RM	reflexive modernization
RSL	registered social landlords
SD	sustainable development
SDP	Social Democratic Party of Germany
SEA	Strategic environment assessment
SOCOG	Sydney Organizing Committee for the Olympic Games

TELCO	The East London Community Organization
TSP	total suspended particulates
TWA	Tibetan Women's Association
UNEP	United Nations Environmental Programme
UNFCCC	United Nations Framework Convention on Climate Change
UNSD	United Nations Statistics Division
WCED	World Committee for Environment and Development
WHO	World Health Organization
WTO	World Trade Organization
WWF	Worldwide Fund for Nature
YPEHODE	Ministry for the Environment, Planning and Public Works
YPEKA	Ministry for the Environment, Energy and Climate Change

1
Introduction

Olympics and the environment?

One of the questions that sprang to mind as I organized this introduction was related to the people who will decide to purchase this book. I realized that, unless the book was to be shelved alongside other books in the social sciences, prospective readers would be unlikely to guess the disciplinary approach adopted here. Even if it were shelved in such a unit, readers might still be surprised by its two themes: the Olympics and the environment. When people think and talk about sport events of great magnitude like the Olympic Games, they tend to think of their environmental dimension in highly restricted ways that don't move beyond the beautification and restructuring of the Olympic host city.

Does this generalization cover those who are solely concerned about environmental issues? The truth is that environmentalists have shown an avid interest in each of the phases that comprise the evolution of sporting events, from their inception as ideas to their deliveries and legacies. Because of this interest, they have often acted as indefatigable monitors of the environmental dimension of sporting events and, in some of the most recent sport mega-events, in particular the Olympics, they have served as important advisors and facilitators.

Today, sport mega-events make an immense effort to showcase their environmental credentials (Hayes and Karamichas, 2012a, pp. 8–14). The core aim of this book is to compare and contrast the environmental legacies that have been bequeathed to four recent Olympic host cities and nations, beginning with Sydney in 2000 (host of the

1

first Green Summer Olympics) and culminating with London in 2012, via Athens in 2004 and Beijing in 2008. This overall objective implies two related goals. First, this book will showcase the valuable role that environmental sociology can play toward critically examining the interaction between the 'social' and the 'natural,' as well as how this examination can complement insights offered by the green sociological imagination. For the purposes of the current study, the 'social' is composed of all of the issues surrounding the running of an Olympic Games, and the 'natural' is composed of all issues that are directly related to the environmental legacies of the Games. Second, the book will offer a detailed appraisal of the extent to which hosting an Olympic Games leads to the Ecological Modernization (EM) of the host nation, an outcome that is strongly promoted by the International Olympic Committee (IOC).

The idea of an environmental legacy is all encompassing. It includes the impact of hosting the Olympic Games on the EM capacity of the host nations. Since at the time of writing the London Games have not yet taken place, the environmental legacy of London 2012 cannot be assessed by means of the vigorous examinations which the other three cases have been subjected to. Despite this issue, this book is not restricted to coverage of existing EM capacities in previous host nations and cities but also offers projections about the environmental legacy of the 2012 Games. Given this projection, it has been necessary to take into account the impact of the financial crisis that the UK (along with the rest of the world) has faced since 2008.

Given the complexity of recent economic history, the reader probably wonders whether the global financial crisis warrants the re-examination of the EM capacities of the other three Olympic host nations and cities. Indeed, the case of Greece, the host of Athens 2004, makes plain the need for such re-examination. In my earlier study of the Sydney 2000 and Athens 2004 Games (Karamichas, 2012a), I concluded that the nature of political parties in government was the single most influential determinant of the post-Olympics EM capacity of the host nation. This argument was substantiated by using examples of specific changes made by labour- and socialist-led governments in both countries. In particular, an electoral victory by the Greek socialist party (PASOK) in 2009 was heralded as the first serious sign that the country was moving toward fully embracing the EM dynamic. Since that time, however, the economic slump has caused that country to experience an unprecedented rate of change in both its social and economic fabric. This has had an immense impact on the EM projections that were made back in 2009. Chapter 7 provides a detailed examination of the Greek case in connection to these developments.

Meanwhile, as a prelude to how these cases and the global economic crisis have been brought together within this book, I want to bring in two descriptors of the Greek crisis that have acted as trigger points for facilitating the discussion of the EM capacities of Olympic host nations and cities. The first descriptor – that the Greek crisis is a 'blessing in disguise' – was taken from a series of articles in *Time Magazine* (Tsiantar, 2010; Crumley, 2010). The second descriptor – that there is a 'canary in the coalmine' – was fished from an article in the *Guardian* (Wachman, 2010).

The first descriptor suggests that the woes that are being experienced and the sacrifices that are being made by the Greek people in attempting to bring their country out of its seriously disadvantageous position may also be in some ways beneficial. In what way and for whose benefit can this crisis be beneficial? This is the question that confronts us. Elsewhere (Karamichas, 2012a, p. 170), I claim that 'the perpetuation of severe structural deficiencies (e.g. the continuation of nepotism and clientelism in Greece) has severely undermined the necessary public legitimacy of the state apparatus to develop its EM capacity'. If the crisis leads to the addressing of these issues by the Greek nation, that development would be beneficial for both the Greek public in general and for business interests in particular. The expectation is that such changes would facilitate *both* the economic and ecological modernization (EM) of the country.

Although the second descriptor – the idea of the 'canary in the coalmine' – does not overlook the peculiarities of the Greek context, it moves beyond them to link the Greek crisis to a general spread of economic crises in other national contexts. The underlying idea here is that the impact of austerity measures in Greece may be replicated elsewhere in the world. This would create adverse effects on the provisions of public services and would lead to a rise in civil contestation. These facts are pertinent to this study in two ways. First, they make a case for examining the impact of the crisis on the EM legacies of hosting. Second, the positive projections that were made about Greece after the election of Papandreou's PASOK in 2009 and were subsequently tarnished with acceleration of the crisis are mirrored in the UK case, for which positive projections have also been made (see Karamichas, 2012b).

To summarize, the onset of the economic crisis and the specific global and national impacts that accompanied it have added an intervening variable in the overall design of this study. Despite this situation, its overall perspective has remained much the same. In essence, this book is a comparative, cross-national study that makes extensive use of the perspectives offered by environmental sociology. In that way, it showcases the scientific and analytical vigour of this sub-discipline of sociology. As we are reminded by Preuss (2004, p. 2), '[i]n order to make a successful comparison of host countries, it is necessary to homogenize the calculation methods employed to determine the final balance'. As such, the contribution that reflexive modernization (hereafter, RM or reflexivity) has made to the discipline of environmental sociology (as that contribution has been expressed through the lens of EM) has been employed in identifying the study's key indicators (Chapter 5), which were then subsequently used in both Chapters 7 and 8 in order to strengthen this cross-national investigation.

The ancient Olympics

A separate issue that has arisen in discussion about this work is the extent to which my treatment of these issues will deal with the ancient Olympics. Most people know that the origins of the Games can be traced back to ancient Greece. The fact that I am a Greek national may have played a role in the questions of my interlocutors. Still, I am slightly perplexed by this interest in the ancient Games. My study is inspired by what, in my opinion, are the most serious issues of our time – the environmental problematic in general and climate change in particular – and by my interest in the relationship between these issues and a quintessentially modern sporting event. My intention is to make use of sociological insights in both my methodology and analysis.

There is already a plethora of publications on the ancient Games by people who are dedicated to the study of ancient Greece. As such, my contribution to this topic would have been of small value. In addition, the relationship of the ancients to the natural environment, as well as the link between the Olympic Games and the concept of environmental protection, are rarely, if ever, at the top of the list of things to learn about the ancients. This is not to say that there are not many things to be said in relation to these issues. For this reason, I want to say a few words about the place of the environment in the social forms of the ancient world.

The ancient Greeks and the environment

Pollution and environmental contamination were not unknown to the ancient Greeks. For instance, in the ancient world coal-driven processes of heating polluted the atmosphere, and lead, mercury and other toxic chemicals from mines were understood to be a threat to public health, and this at a time during which there were no provisions for the storage and treatment of such waste products. At the same time, the ancient Greeks made provisions for the transfer of human waste to places of considerable distance from the walls of the city, as well as for its reuse in the fertilization of farming plots. Moreover, as early as 478 BC, we can find the first expressions of concern about the population and spatial growth of urban centres in ancient Greece (Gouta, 2009, p. 10).

For the ancient Greeks, this concern amounted to the potential reduction of public space. Following Hannah Arendt (1958), we can say that this process constitutes the pure essence of democratic socialization. Georgi (2010, p. 87) has summed up Arendt's rationale:

> In ancient Greece, the 'public' was defined as something which had the possibility of wider publicity, what could be seen and heard by everybody, and which also had, because of its communicative breadth, intense political tinge. While on the contrary, the 'private' in the ancient Greek world, meant deprivation from the objective societal links, in essence deprivation from real relationships for development of human existence. As such, the 'private' was characterised by an association with the negative.

Interestingly enough, in her attempt to exemplify these points, Georgi (2010) breaks down the concept of 'public space' into three different categories: squares, gardens and sacred thickets. Considering that sacred thickets can be understood as the predecessors of today's public parks, Georgi makes a statement that is of considerable importance for any attempt to assess environmental factors within the mental framework that guided the originators of the Olympic Games. In 'other civilizations, Roman times, Renaissance etc.,' she writes, 'green space was only composed by the gardens of Royal Palaces, which had some public character' (ibid., p. 87). When we examine the actual public spaces that inspired Arendt to make her claim, we discover that the ancient Athenian *agora* (market) was initially a square-like structure that 'operated as a symbol of free expression and distribution of ideas and goods' (ibid., p. 88). Furthermore, it is worth mentioning that these ancient sacred thickets were protected by laws that prohibited the pasturage of animals, the removal of dry tree branches, and tree felling. In addition, the golden age of Pericles was marked by the construction of the first public gardens in Athens. These gardens 'constituted notable meeting points between philosophers and their students who were taking a stroll whilst discussing in walking paths shaded by planes and acacias. In general, the organization of free public spaces in Greece was intimately connected to religious worship since every one of the Gods had a tree dedicated to him/her that was seen as sacred. In this context, the pine was dedicated to Pan, oak to Zeus, beech to Hercules, holm oak and poppy to Mars, mint to Pluto, and myrtle and rosebush to Aphrodite' (ibid., pp. 91–2). According to Elliot and Thomas (2009, p. 1),

> Greek religion recognised humanity's oneness with nature. The ancient Greeks could not conceive of a sacred area without trees. [For instance,] on the rocky summit of the Acropolis when the Parthenon was built, holes were excavated in the solid rock and trees were planted, in rows, to flank the temple. [In addition, sacred] groves of trees were preserved and hunting was forbidden in these groves. While it is stretching the point to see these as national parks in anything like the modern sense, it is clear that the Greeks had a respect for nature.

The Romans changed this attitude through more pragmatic and utilitarian motivations that sanctioned the exploitation of nature. This Roman perspective has been bequeathed to us. In this perspective, we can detect the rationale that has underpinned industrialization. There is little doubt that the modern Olympics have been permeated almost exclusively with this mentality. This is an issue that is developed in the chapters that follow. In summary, however, there is little reason to doubt that the ancient Games were held in a way that was harmonious with the natural environment. After all, the scale of the event in ancient times, irrespective of its wide connection to the Hellenic World, was local and nothing like Games' contemporary global scale. Substantial challenges have been posed to traditional understandings of the Games. These challenges include a range of publications that came out in time for the Athens 2004 (see, for instance, Katsikas and Nikolaidou, 2003; Perrottet, 2004; and Spivey, 2004). Those Games led Kalogeratou (2007) to recontextualize Baudrillard's statements on the eve of the first Gulf War (Baudrillard, 1995) and ask, 'did the 2004 Olympics take place?' Kalogeratou did not dispute the fact that an event called 'the Athens 2004 Olympics' took place. She did, however, question the extent to which that event was an Olympic Games according to the 'traditional cultural characteristics that define' those Games, a claim made along the same lines that the French philosopher questioned the extent to which the remote-controlled missiles used in the military operation of the first Gulf War signified a 'real war' (Karamichas, 2012c). Kalogeratou writes:

> We often ignore how much the contemporary staging of Olympic Games has been differentiated, from antiquity to nowadays but also from the first modern Olympics to their most recent edition. Many of those elements that nowadays constitute characteristics of authenticity are either fictitious or counterfeiting (such as the Marathon and the Torch relays), whilst many of their ancient characteristics would have caused outcry nowadays (for instance the participation in the sports only by men, Greek citizens who competed naked).
>
> (Kalogeratou, 2007, p. 161)

In keeping with this interesting subject, I expand further upon the issue of continuity between the ancient and modern Games in Chapter 2.

I want to move forward now and elaborate upon the actual themes of this book. I will start by highlighting its overall concept, the Olympics and the environment, and go on to discuss how the disciplinary specialization of environmental sociology may be of interest.

Sociology, the Olympics and the environment

The study of sporting events is generally seen as something outside the sphere of sociological endeavour, (see, for instance, Roche, 2000, p. ix). The study of sporting events has not generally been taken seriously by the discipline. As Bourdieu (1990, p. 156) writes, 'the sociology of sport: it is disdained by sociologists, and despised by sportspeople'. If one adds to this the environmental dimension of these events, this disdain becomes even more acute.

There are concerns, of course, about the capacity of the 'sociological imagination' to show its relevance to life outside of academia, and these concerns escalate if one takes into account articles advocating the need for a sub-discipline of environmental sociology. Furthermore, attempts to ascribe sociological connections to the environmental problematic still raise eyebrows in certain sociology departments. As many environmental sociologists can confirm, Durkheim's 'social facts injunction' (Catton, 2002) is to a certain extent responsible for the exclusion of environmental concerns from the discipline. The truth is that even today – more than a decade after the publication of Raymond Murphy's 'Sociology as If Nature Did Not Matter' (1995) – it is not exaggerating to say that, irrespective of some notable improvements, the discipline of sociology is still largely conditioned by a subscription to the 'human exemptionalism paradigm' (HEP).

Despite this exclusionary tendency and with both climate change having become unequivocally the most valent issue of our times and the London 2012 Olympics approaching, British sociologists have had to take an active role in producing a critical and constructive perspective about sport and the environment. This new perspective is evident in the most recent work of Anthony Giddens on climate change, *The Politics of Climate Change* (2009), as well as the addition that he made to the online version of his popular sociology textbook, *Sociology*, which includes a new chapter title 'Sociology, Sport and the Olympics' (Giddens, 2008). Given this situation, the amalgamation of sociology and components from the natural sciences may become a self-evident and necessary development. Similarly, sports studies may be infused with environmental concerns as more and more sporting events subscribe to environmental management systems and attempt to account for their carbon footprints.

Modernity and modernization

Sociology is unequivocally a product of modernity. Chapter 4 argues that because modernity was intimately interwoven with the existence rational reasoning in the Enlightenment and the unparalleled technological and scientific developments of the time, environmental considerations were excluded from the origins of sociology as a discipline.

What is the relationship between modernity and the Olympics as a sport mega-event? I touched briefly on this issue earlier in my rebuke of the generally imaginative associations made between the ancient Olympics and the contemporary Games, and Roche (2000, pp. 7–8) offers an important note about this relationship:

> Interpreting their more general significance for modernity we can observe that international mega-events like the Olympic Games – along with the event-based aspects of other genres in contemporary popular culture, such as film, pop music, professional sport, and so on – offer national and international publics the experience of something unique, dramatic and literally extra-ordinary. That is, they promise modernity the occurrence (and, ironically, also the control) of charisma and aura in a world often appearing as excessively rationalistic and as lacking any dimensions beyond the everyday lifeworld and its mundanity. Also, in their calendar and cycles mega-events offer modernity a vision of predictability and control over time, over the pace and direction of change, in a world where social, technological, ecological and other changes can often appear "out of control" – unpredictable, too fast to adapt to, purposeless and generally anomic.

A relatively more recent expression of the nature of this modernity has been seen in the appearance and development of the modernization perspective. The modernization school made its entrance in both academic and policy debates with the post-World War II rise of the US as a world superpower. The extreme devastation that Europe had experienced during the war was coupled with the onset of the Cold War and the disintegration of European colonial empires in Asia, Africa, and Latin America. For So (1990, p. 17), emergent, post-colonial 'nation-states were in search of a model of development to promote their economy and to enhance their political independence. In such historical context, it was natural that American political elites encouraged their social scientists to study the Third World nation-states, to promote economic development and political stability in the Third World, so as to avoid losing the new states to the Soviet communist bloc'. The modernization school thus created a single view about the development process in which the economic and the political were seen as two mutually reinforcing components. This view, which aimed to put poorer countries on the road to Western-style development and progress, was challenged by the radical dependency school during the late 1960s and by another, alternative view offered by the world-system school during the late 1970s (ibid., p. 12). When these three dominant schools appeared to converge in the late 1980s because of the many significant changes in environmental policy that were taking place in many OECD countries, the EM perspective first emerged and gained 'growing popularity [...] in part from the suggestive power of its combined appeal to notions of development and modernity, and to ecological critique' (Christoff, 1996, p. 476). Chapter 5 shows how Christoff (1996) has demonstrated the existence of different EM variants. This is particularly important in appraising the precise character of EM that may be facilitated by Olympic Games hosting.

The hosts of Olympic Games and the environment

As I said at the outset, the connections that are usually made between hosting the Games and the environmental effects of this hosting are mostly thought about in relation to regeneration of the host city in general and to the designated area of the Olympic facilities in particular (e.g. Sydney Harbour in the case of Sydney 2000; the Lower Lea Valley/East London in the case of London 2012). Regeneration is without a doubt one way in which an environmental contribution can be made by hosting the Olympic Games. But there is another, more important connection with which this book engages and that is what I have called the ecological modernization (EM) of the host nation. The importance of this modernization is supported by the emphasis that the IOC gives to the environmental impact and legacy of Games in terms of the EM perspective. Preparations to meet this proclaimed ambition on the part of host nations require coordination among different state bodies, engagement with civic organizations, and the restructuring of host cities' infrastructure – what Rutheiser (1996) has called 'imagineering'. Within this framework, the dominant view is that hosting the Olympic Games goes beyond the successful preparation and delivery of a one-time event and instead has a long-term impact on the host nation's capacity for EM (Weidner, 2002). Elsewhere, I have argued that 'we can see a process akin to *engrenage* [...], in similar terms conceived for policy-making in the nascent European Community by Jean Monnet, in that the process of meeting the IOC's environmental standards could both drag with it the host nation's institutional framework and set a precedent that other nations would strive to emulate' (Karamichas, 2012a, p. 156).

The engrenage dynamic of hosting the Olympic Games

It is worth noting that Laurence Nield (TEE 2009) has argued that hosting the Olympic Games is an important factor in the modernization of the host city as 'it can improve the city's institutional framework as well as bring substantial economic development, new offices, new residences, better environment'. Chapter 7 puts under the microscope the legacies of Sydney 2000 and Athens 2004 in relation both to the modernization of the two cities and to the greater EM capacity of Australia and Greece. Furthermore, we should not lose sight of the fact that the 1992 summer Olympics in Barcelona, which have been seen as a paradigmatic case for the rejuvenation legacy of hosting the Games (see Brunet, 2009), 'resulted in environmental damage in a number of locations, including the natural park of Collserola, the last remaining habitat for a number of plants and animal species' (Hannigan, 1995, p. 29).

The idea of *engrenage* is also evident in relation to other qualitative issues like human rights. For instance, David Close and his colleagues see the Olympics as a springboard for China's 'coming of age more broadly ... through its greater compliance with the Olympic movement [...] and connectedly, with the values enshrined in universal human rights'. (Close et al., 2007, p. 39). And we should not forget the precedent of the 1988 Seoul Games. South Korea was at the time under political leadership that did not appear to have completely dissociated itself from the military structure that it had replaced, and it was still technically at war with the Democratic People's Republic of Korea (DPRK). Nevertheless, 'the new spotlight shining on South Korea in the period leading up to the Seoul Games had an important effect on the speed with which the South Korean leadership was willing to move to the direction of political reform' (Pound, 2008, p. 91). Finally, Atlanta invested a great deal of effort in representing itself as an ideal promoter of the human rights ideals of the Olympic movement 'through its claim to being the international human rights capital of the world' (Rutheiser, 1996, p. 229), yet the aforementioned 'Imagineering' of Atlanta was implemented in a way that minimized any possible opposition to projects that were bound to lead to mass displacements of people (Rutheiser, 1999, p. 332).

Research questions

In order to fully understand the place of the environment in hosting the Olympic Games, this study primarily employs the insights of environmental sociology in general and EM in particular. Since four different host cities have been used in this examination, this qualifies as a cross-national study. Like Burbank et al. (2001), the approach adopted here can be described 'as a focused comparison of case studies' (George and McKeown, 1985). The four host cities under investigation differ in many respects, although very broadly speaking all have industrial capacities as well as access to modern technology. Still, these superficial similarities should not disguise the vast differences that exist between them and the nations that they represent. Australia and the UK are both advanced industrialized nations. The latter, in particular, is a member of the G8, and London is at the epicentre of global capital flows. The differences with Greece, however, are vast. Greece is a fully democratized country, an EU member state, and the most developed country in the Balkans. However, the modernization project of Greece is still infused with a number of structural deficiencies that make any leap in its capacity for EM rather uncertain. The severe economic situation in which the country found itself in 2010 brought these impediments to the fore. Interestingly, cultural similarities exist with the Chinese case, in particular with the role played by personal connections and the familial in policy-making contexts. As will be demonstrated, however, the People's Republic of China (PRC) has fully embraced the modernization process since the 1970s. For that reason, it is not much of a surprise to find that significant progress has been made in China on the environmental sustainability front. Another important point of difference that is intimately connected to the main research objective of this book is that, of the four host nations under examination, the UK stands out as the nation with the most established environmental credentials. Yet these credentials have been confronted by the impact of austerity measures undertaken by the current conservative-liberal coalition government. For that reason, as has been already indicated, the economic crisis and its accompanying austerity measures have been used here as an intervening variable.

Key themes

The discussion has been organized through the following interrelated parameters:

- Sociology, modernity, and the environment;
- The inclusion of the environmental parameter in the policies of the IOC, in particular with its becoming the third pillar of the Olympic chapter after sport and culture;
- The place of the environment in the bidding applications of prospective host cities;
- The impact of mega-event hosting on infrastructural planning of the city and associated changes in transport networks; and
- The facilitation of environmental-capacity building in the host nation or Olympic Games host as an opportunity for the EM of that nation.

Outline

Chapter 2 highlights the multifaceted character of the Olympic Games, which simultaneously exhibit modern/non-modern, local/national, and global/non-global dimensions (Roche, 2000). I begin by discussing the connections of the Games with those of the past through pageantry, ceremonialism, and the ideals of athletics with which they have been associated since their revival at the end of the nineteenth century. I then chronicle the association made between the Games and economic development. In particular, I describe how the modern

Olympics were transformed from an elite athletic competition into a more popular event that was in need of significant financial input from corporate actors. This factor has conditioned prospective host nations to showcase investment opportunities in order to take advantage of global capital flows. In this context, preparation for the Games leads to significant alterations to existing urban environments that may lead to social confrontation, while the strict deadlines assigned to Olympic-related projects often lead to the bypassing of or complete disregard for existing legal frameworks and public accountability (see Andranovich at al., 2001; Burbank et al., 2001; Hiller, 2000). As such, the Olympic Games experience the twofold dynamic of support and opposition that is attached to all other modern/modernizing projects. The second chapter closes by arguing that the environmental factor made a belated entry into this dynamic and that this emergence was very much in tandem with the emergence of environmental concern among the publics of economically advanced nations. Despite this concern, however, the adoption of environmental issues by the IOC took a great deal of time and its effectiveness is still a contentious issue.

Chapter 3 offers an overview of environmental concerns around the world. Its main focus is the development of environmentalism since the 1960s and 1970s and the various discourses that have supported it. In order to substantiate the importance of this point in the developmental sequence of environmental concern, I have started the discussion by providing an account of various manifestations of environmental concern within the ancient world. I subsequently move toward an account of the foundation of the first nature protection/conservationist organizations during the nineteenth century. In this way, there is no doubt about the challenging perspective that marked the new wave of environmental concern of the late 1970s. This perspective is further illustrated by paying attention to a variety of influential ideological currents around environmentalism through the mid-1980s and early 1990s, when the first signs of a new, and less anti-systemic, perspective made their appearance. This chapter concludes by arguing that this point is precisely where we can see the first signs of interest in the environmental issue on the part of the IOC in their requirements for awarding the Games.

A significant objective to this volume was to demonstrate the ability of the sociological discipline to engage with environmental issues.

For this reason, Chapter 4 identifies extensive connections between the emergence of environmental concerns and the incorporation, during the same period (the 1970s), of the environmental issue into sociological endeavour. The latter event was manifested in the development of a new sociological sub-discipline, environmental sociology. This chapter embarks upon a detailed exploration of the issues, debates, and conflicts that have marked the perspectives put forward by the first environmental sociologists. The chapter offers a fairly detailed account of the actual engagement of classical sociology – in particular, of the representative figures of Durkheim, Marx, and Weber – with the environmental factor. In doing so, this chapter claims that, not only did these founding fathers not completely disregard the biological, but that in certain cases used the 'natural' as an essential component of their theorization. This account is followed by a description of scholarly works that make good use of these classic perspectives in their examinations of the environmental problematic.

Chapter 5 follows up the preceding discussion by exploring one theoretical perspective that managed to simultaneously give new momentum to environmental sociology and establish the environment as an important item of sociological engagement. This is the reflexivity perspective. The reflexivity perspective first emerged and made inroads in work produced by Ulrich Beck and Anthony Giddens. This work argued that reflexivity is composed of two facets: Risk Society and Ecological Modernization (EM). After discussing both reflexivity facets in detail, the practical, policy-based implementation of the latter facet is further described in order to lay the foundations for a cross-national examination of the legacies of Olympic Games hosting on the environmental front.

Chapter 6 provides an overview of connections made between environmental issues and the organization and planning of the Olympic Games. This section essentially follows the treatment of the rise of environmental concern among Western publics dating back to the 1960s and 1970s, and it puts forward the claim that the IOC has been extremely slow in adapting its procedures to factor in these concerns. Since the addition of environmental protection to the Olympic Charter in 1996 and the development of the IOC's very own Local Agenda 21 (LA21), the incorporation of the environmental factors into planning, organizational, and legacy concerns has been constantly evolving. This chapter closes by providing coverage of the latest developments on this front.

Chapter 7 is dedicated to providing a thorough answer to the question at the centre of this book: 'To what extent does hosting the Olympic Games lead to the EM of the host nation?' The case studies used to answer this question are those of the Sydney, Athens, and Beijing Olympics, and their examination is closely tied to a study of the different phases that comprised their development. This approach closely follows the developmental sequence of the mega-event as that concept was put forward by Hiller (2000, p. 192). For the purposes of this study, the concept was adjusted to include the 'pre-event' phase of IOC bidding applications and preparations toward the fulfillment of environmental commitments made by successful bidders, the 'event' phase itself, and a 'post-event' commitment to environmental sustainability (ES).

Chapter 8 is dedicated to evaluating the prospects of London 2012 in terms of making an effective contribution to the existing EM capacity of the UK. This exploration follows the same steps of analysis that were adopted throughout Chapter 7, but like the Greek case, it discusses the impact of the austerity cuts. The riots of August 2011 are also discussed because of their close relation to the Olympic boroughs.

Chapter 9 pulls together all of these threads and discusses the relevant lessons learned for the sociological examination of the environment in future Olympic Games.

2

The Olympic Games: A Quintessentially Modern Project

A product of modernity

The position that the Olympic Games are actually a product of modernity is likely to be met with outcries from some sectors. 'The origins of the Games are traced to ancient Greece!' such people say in evident fury. It can be argued, however, that these same people are attacking the current institution of the Olympics as well as the culture for which they claim to act in defence. In thinking about the connection of the modern Games to their ancient past, we have to examine the extent to which we are referring to the same event. For many people, this will appear to be a pointless exercise, but it is a must if we are to properly understand our subject. A less pointed question might be: 'To what extent are the Greeks of the past and the Greeks of the present the same people?' This question has been asked innumerable times and for various reasons in relation to the inhabitants of the southern Balkans, who have identified themselves with that name since the foundation of the Greek state in the nineteenth century. What is crucial is that the Greek public appears to hold an unshakable conviction about the uninterrupted continuity of the Greek nation across the centuries and as such clings to the belief that it holds inalienable rights over Olympism. This phenomenon is well known, and it has been encouraged and exploited by both internal and external actors. Let's examine this issue in more detail.

17

Demystifying the origins of the Games

The ancient Olympic Games occurred 293 times and lasted for twelve centuries (1169 years). They were abolished in 393 AD by a decree of Emperor Theodosius, who had made Christianity the official religion of the Roman Empire (both East and West) and saw the Games as an expression of idolatry. (At the time, they were held in Zeus's honour.)

After this decree, many attempts to revive what was the most important athletic and spiritual event of antiquity were attempted without much success. Indeed, many cultural initiatives and competitions, all referred to as the 'Olympics', took place in various European countries (including England, France, Sweden, Canada, Germany and Greece) during the eighteenth and nineteenth centuries (Kidd 2005, p. 145). Kritikopoulou (2007, pp. 51–2) writes:

> At the end, the dream of resurrecting the Games became a reality by Baron Pierre de Coubertin [1863–1937], [...] when in Autumn 1892 he made an official proposal for the resurrection of the Olympic Games at the amphitheatre of Sorbonne to an audience of historians and representatives of all sports. In 1894 in an international conference that took place in the Sorbonne, with the presence of 2000 country representatives from around the world, Coubertin's proposal was unanimously approved and Athens was selected to conduct the first modern Olympic Games in 1896, namely 1503 years after their suppression. Coubertin was assisted in his effort by the Greek poet and writer Demetrios Vikelas, who became the first President of the International Olympic Committee.

Young (2005, p. 13) has unearthed the following interesting details about the appointment of Vikelas:

> At first, Vikelas did not support Coubertin's nomination of Greece, but later the same evening he changed his mind, and 4 days later [...], Vikelas himself made a second more formal and far more successful proposal for Athens' candidacy at a plenary session. In the meantime, he had communicated with Athens by telegram. Vikelas' Athens proposal was approved by acclamation.

A compelling story has also been told in relation to Pierre de Coubertin, which diverts from the hagiographies that Kritikopoulou appears to have followed. For instance, according to Boykoff (2011, pp. 42–3), de Coubertin was 'an eccentric Anglophile who saw in the sporting culture of Thomas Arnold's Rugby School the magic formula for Britain's imperial dominance. Here, in the mix of rigorous discipline with manly self-display, lay the means to reinvigorate the French nation after the humiliation of the Franco-Prussian War'. For de Courbertin, 'the games curricula of the all-male British upper-class schools provided the best means of providing for physical education', something which the French youth was severely in need of (Kidd, 2005, p. 145). Interestingly, de Coubertin's original vision for the revived Olympics included the promotion of international peace in an age of rampant nationalism (Hoberman, 2004). Likewise, it is important to remember that de Coubertin's ambitions occurred at a time that Robertson (1992) has described as the 'take-off face of globalization', a period

> during which organisations like the International Red Cross and the Boy Scouts were created, a global language (*Esperanto*) was invented, global prizes (e.g. Nobel Prizes) were initiated, and cities held the world fairs. Although he started his journey in order to improve the fitness of French youth, he came to see the value of sport from an increasingly internationalist perspective. It was only in 1890, when he visited Dr. W. P. Brookes, the founder of the Wenlock Olympic Games in England, that de Coubertin came to see modern Olympics as the vehicle for the international athletic that he was increasingly advocating.
>
> (Kidd, 2005, pp. 145–6)

The archaeological exploration of Olympia and the display of a model of the stadium at the Paris World's Fair of 1890 gave de Coubertin the opportunity to graft 'his ideas about sport and education to the Olympics of antiquity'. This 'proved a genius stroke of public relations' as it offered the Games a much needed historical verification for their survival during their inaugural century (ibid. p. 146).

In agreement with the point made by Kalogeratou (2007) about the similarities between the ancient and modern Olympics, Kidd has continued to further demystify the connection of the modern Games to the ancient Olympics and the ways that an ideal image of the latter has been carried forward in service of particular political goals. Kidd points out that

> [t]oday's ideals of sportsmanship and fair play would have been laughable in the ancient Game. For instance, the popular combative events were conducted with little concern for safety or fairness. There was no equivalent to the modern eight categories to balance size and strength. Bouts were essentially fight-to-the-finish, and the best athletes tried to psyche out their opponents so that there was no competition at all. The intrinsic value of participation, the pursuit of personal bests and records, and the congratulations of opponents – all familiar values and rituals in the modern Games – would have been meaningless to them. Winning was the only goal.
>
> (Kidd, 2005, p. 146)

This deconstruction of the Olympic ideal, although partly recognized by other scholars (see, for instance, Kritikopoulou, 2007, p. 51) is shocking to those of us who have learned to associate the ancient Games with the 'cultivation of an athletic ideal characterised by noble rivalry and not competition' (ibid. p. 52). The presence of that understanding has led Kritikopoulou to make the following statement, which, while somewhat exaggerated, is nevertheless representative of the feelings and ideas that sprang out of the modern construction of an ancient Olympic ideal:

> In the current times, which are marked by social and economic transformations that compose the functioning of a globalised society, there is a diversion from the original institutional pursuits through its promotion as a commercial event. Even if that can be justified and partly rationalised in the context of evolutionary development, it cannot deter us from our value commitment in favour of the management of the Olympic Games institution not only as an event of commercial or athletic but also as a means of moral and intellectual progress and rejuvenation'.
>
> (Ibid, p. 53)

A further major blow to this idealistic perspective is dealt by Kidd (2005, p. 146), who argues that the 'ancient Greeks did not believe that the Games could end military conflict. The truce was designed merely to prevent wars from disrupting the Games. In fact, the ancient Games developed *out of war* games' (italics in original).

We can further confirm this perspective if we think about the Olympic torch relay that follows the ceremonial lighting of the torch in Olympia. The truce and the Olympic torch relay are 'carefully orchestrated myths which, through repetition, have become established systems of meaning' (Hayes and Karamichas, 2012a, p. 6). Indeed, another fact that is conveniently left out of popular Olympic historiography is that Joseph Goebbels, the Nazi minister of propaganda, came up with the Olympic torch ritual for the 1936 Berlin Games (Krüger 2004, pp. 45–6).

It is also worth noting that the aforementioned perception of an original Olympic truce was 'revived' after a successful campaign in the 1990s (Briggs et al., 2004). It is interesting to see how Payne (2006, pp. 273–4) has narrated the events surrounding that campaign:

> The Greeks, never short of ideas, were soon petitioning the IOC to create a permanent centre to promote the idea of the truce, as another Olympic legacy of their Games. George Papandreou, the eloquent Greek Foreign Minister, championed the cause and eventually succeeded in persuading a reluctant IOC to sign up.

Payne's little history provides us with additional insight into the vision that has characterized the policies and politics of Papandreou and which can help us to add to our understanding of his handling of the post-Olympics legacy in Greece at a time of immense financial crisis.

Regardless, this historic trivia about the Olympics cannot lead to a rejection of the potential that can be accrued to symbolic acts, like the Olympic flame and the related torch relay for the promotion of various issues (peace; human rights, democracy) (Hayes and Karamichas, ibid. p. 6).

In this light, we might even justify the constructive potential that Kritikopoulou (2007) saw. There are two important elements to these histories that are worthy of further consideration: the place that modern Greece holds in the pageantry and ceremonialism of the modern Games and how this activity is used by disparate sources; and the fact that the Olympic Games should be seen as quintessentially modern.

The honorary first place of entrance for the Greek flag and team was established during the Amsterdam 1923 Olympics (Kritikopoulou 2007, p. 74). This ceremony is one of the elements that have added to the construction of the modern Greek nation in the public imagination. Yet the idea of the uninterrupted continuity of the Greek people since ancient times has, on various occasions, been emphasized in order to support highly nationalistic claims. Elsewhere (Karamichas, 2012c, p. 165), I have suggested that this kind of transaction is one explanation for the continuous attempts on the part of the Greek nation to host the modern Olympics. Kidd offers the following argument, which was originally put forward by the classical scholars Finley and Pleket (1976, p. 132):

> What we choose to think about sport in the modern world, in sum, has to be worked out and defended from modern values and modern conditions. Harking back to the ancient Greek Olympics has produced both bad history and bad arguments. It might be right or wrong to forbid Olympic athletes to profit financially from their medals (as one example) but the answer will not be found in the northeastern corner of the Peloponnese, and surely not when what happened there two thousand years ago is distorted and perverted to fit one or another modern ideology.

Olympic Games as mega-events

Mega-events must be understood in the context of the mega machine of Lewis Mumford (1966). In Mumford's words, '[t]he habit of "thinking big" was introduced with the first human machine; for a super-human scale in the individual structures magnified the sovereign authority. At the same time it tended to reduce the apparent size and importance of all the necessary human components, except the energizing and polarizing central element, the king himself' (Mumford, 1966, p. 202). This perspective has rarely been taken into consideration in scholarly works that deal with notions of mega-projects or mega-events, notwithstanding what might be understood as mega-machine offshoots, such as capitalism, industrialization, modernity, and the Veblenian 'conspicuous consumption' and waste (see Veblen, 1899/2005).

Although at this time many of the mega-edifices that result from such mega-events are ephemeral and conditioned by time–space constraints rather than wishes for perpetuity, the legacy of *hosting* the mega-event has been rhetorically mobilized in attempts to create consensus among the general public. Horne (2012, p. 38) argues that 'legacy is a warm word, sounding positive, whereas if we consider the word "outcomes" it is a more neutral word, permitting the discovery of both negative and positive outcomes'. Olympic legacies can take various forms and guises, including appeals to national or city pride that sovereigns of the past may also have used. Flyvberg et al. (2003) use the example of mega-projects – the edifices of modernity – some of which are intimately connected to mega-event hosting, such as the Channel tunnel for the London 2012 Olympics. The Channel tunnel is of particular interest because its construction attempted to make the lowest possible contribution to increases in greenhouse gases (GHG) emissions. Flyvbjerg et al. (ibid., p. 2) claim that '[m]egaprojects form part of a remarkably coherent story,' and they continue by providing their readers with a tour of various perspectives that substantiate this claim, ranging from sociologist Zygmunt Bauman, who sees them as being a key part in what he calls the 'Great War of Independence from Space', to Paul Virilio and others who understand them as a manifestation of the 'end of geography' and the 'death of distance', to Bill Gates, who 'has dubbed the phenomenon "frictionless capitalism" and sees it as a novel stage in capitalist revolution'. There is, however, an inherent paradox that dominates these projects and that bears a striking resemblance to the projects of Mumford's mega machine. As Huber (2003, p. 595) writes,

> Less explicit is another feature, namely that, once being started, these megaprojects cannot be stopped as the economic, political and prestige losses are considered disastrous. Carrying out megaprojects is then described as a combination of international ignorance, megalomania and – in later phases of each project – blame shifting.

Megas and civil society

It is unsurprising that, as far as mega-projects are concerned, '[c]ivil society does not have the same say in this arena of public life as it does in others; citizens are typically kept at a substantial distance from megaproject decision making' (Flyvbjerg et al., p. 5). Democratic accountability is also relatively absent from the planning of mega-events. For instance, Andranovich, Burbank, and Heying (Andranovich et al., 2001; Burbank et al., 2001) have demonstrated in their study of mega-event politics that citizen participation and democratic accountability in decision-making were notoriously absent in three host US cities (Los Angeles, Atlanta and Salt Lake City). This view is strongly supported by Hiller (2000, p. 193):

> [m]ega-event planning is top-down planning. Just as the idea to bid is itself normally an idea of an elite group that then tries to sell the idea to other elites and urban residents at large, so is mega-event planning the specification of a design plan (sometimes with site alternatives) for how the city could accommodate the event to which citizens will be given an opportunity to react. The idea of citizen participation is, then, primarily merely responding to a plan conceived by others.

At the early stages of planning, where everything appears to be only hypothetical, citizen participation is bound to be muted.

There are, of course, cases where a more democratic approach has been successfully implemented, such as Toronto's bid for the 1996 Summer Olympic Games or the 1988 Calgary Olympics (Gursoy and Kendall, 2006, p. 604), but these more democratic approaches resulted in the failure of Toronto's bid and the award of the 1996 Summer Olympics to Atlanta (see Lenskyj, 2000). After all, both mega-projects and mega-events 'have fixed completion dates that must follow a tight schedule which, on the one hand, ensures results rather than unending deliberations but, on the other hand, may produce autocracy against which opposition may arise' (Hiller, 2000, p. 198).

It is important also to note, however, that the democratic deficit that appears to characterize the planning of mega-projects and mega-events has led to important instances of social contestation and protest mobilizations by citizen groups as well as by the more regular corps of social activists. In terms of the larger story of modernity, it has also been recognized that 'the great mega-project era came to an abrupt end during the late 1960s and early 1970s, as new social movements erupted and local residents rose up to defend their neighbourhoods. Numerous new policies and procedures were adopted to safeguard environmental, neighbourhood and preservationist values' (Altshuler and Luberoff, 2003, p. 219).

Still, 'both globalisation and the restructuring of cities have been powerful factors in enhancing the attractiveness of mega-events as stimulants to urban economic development' (Malfas et al., 2004, p. 211; see also Chalkley and Essex, 1999a). In their review of the literature on the Olympic Games and other sporting mega-events, Malfas et al. (2004, p. 218) write:

> Despite the widespread criticisms surrounding the institution of the Olympic Games, which mainly challenge the connection between the ideas of Olympism and the contemporary nature of the event, the Games continuously grow in magnitude and significance. In effect, the contemporary Olympics sustain the status of a mega-event, and economic benefits are the prime motive for all the interests involved in the hosting of the Games, be it the local Government, which seeks urban development of the region through infrastructure made for the staging of the event, or the corporation that becomes a sponsor of the event to attract publicity. While bidders battle for the kudos of winning the hosting of a mega-event, the desired economic, fiscal, social, cultural and political outcomes are expected to justify their actions but further research in the area is necessary to judge the benefits of such undertakings in light of costs and potential negative impacts.

These aspects of mega-events are also developed by Horne (2012 and 2007) and Roche (2006), who argue that mega-events are of paramount importance to the direction that nations take in the context of globalization. As such, mega-events have become a central element of urban modernity, and they exhibit many of the contradictions of what Klein (2007) has called 'disaster capitalism'. For Horne (2012, p. 39), 'the relationship of sports megas to developments in contemporary capitalism is evident', and although the idea that capitalism advances on the back of disasters is not new, 'Klein's book is a valuable insight into recent history'. Horne (ibid., pp. 39–40) is quick to recognize that readers of his summary of Klein's arguments may wonder exactly what the connection is between 'disaster capitalism' and sporting mega-events. He thus writes that 'these sporting spectaculars can be viewed as the twin of disaster capitalism's shock therapy, involving their own shocks and generating their own forms of awe. Winning a bid to host a mega-event, putting the fantasy financial figures of the bid document into operation, dealing with the proposed location before and dealing with it after the event has taken place, are just some of the moments where shock and awe is generated by sports mega-events'.

In agreement with these views, Spilling (1996, p. 323) describes a mega-event as 'an event that generally attracts a large number of people, for instance more than 100,000, involves significant investments and creates a large demand for a range of associated services'. More emphatically, because of the global attention that the host city attracts in the process, the improvement of the city's profile becomes an uncontested necessity. In addition, as Essex and Chalkley (1998, p. 201) remind us, '[i]n the context of post-Fordism, as cities strive to present themselves as consumer, leisure and cultural centres, the growth of the games as an international spectacle has offered its host increasing opportunities to claim physical attributes and images which attest to its distinction, taste and eminence'. Having established the intimate connection between capitalist developments and mega-events, it is crucial to identify the main actors behind the promotion of these developments to the general public in democratic polities.

Caratti and Ferraguto (2012, p. 111) bring attention the fact that '[e]vents are usually developed under a special "regime", a conferred coalition drawn by the private and public sectors, and which is capable of overcoming fragmentations in local policy'. In order to legitimize the importance of hosting these events and that way mobilize the support by the general public, promises of a whole range of different long-term legacies are put into action. According to Girginov and Hills (2010, p. 438) these legacies are 'a means to redirect and expand the growth of the Olympic Games'. 'Whilst legacies are typically tied by hosting coalitions to the staging of the mega-event itself, they respond to longer-term social and political agendas; and – since Sydney hosted the 2000 summer Games – claims from event promoters that their staging of the "greenest" or "most sustainable" event so far will have long-term impacts on social and economic processes and structures' (Hayes and Karamichas, 2012b, pp. 249–50).

Olympic Games and social contestation

Earlier, I mentioned that there have been attempts to facilitate more democratic processes around Olympic bids, primarily in Toronto and Calgary. As Lenskyj (2000, p. 70) has pointed out about the public consultation that took place around Toronto's bid:

> For most of the individuals and groups involved – both supporters and critics – attention focused on the social impact and legacy of the Games, and not simply on the Games as high-performance competition or sporting spectacle. The economic impact was a key concern: supporters portrayed Toronto's hosting of the Games in terms of increased job opportunities, a legacy of housing and recreational facilities, and a healthy financial surplus, while critics pointed to the debt incurred by other host cities and the negative economic impact on already disadvantaged groups such as the homeless and those living in poverty.

In the case of Calgary 1988, however, this more democratic approach did not produce the negative results that marred Toronto's bid. That does not mean that Calgary had escaped the civil protest and contestation that is more often than not attached to hosting the Olympic Games. In characteristic fashion, controversies in Calgary over project locations and impacts were quickly put aside by means of the orchestration of 'a speedy end to community consultation and the democratic process' (ibid., p. 118).

Indeed, protest action has been a reoccurring feature of the Olympic Games since their modern reincarnation in Athens in 1896. Over the decades, this action has taken many forms and it has not always targeted only the projects and developments that are made by prospective or successful host cities. As Cottrell and Nelson (2010, p. 1) have argued in a study of Olympic protest, the Olympics provide 'a unique opportunity structure for a range of actors to exercise power in pursuit of their goals'. These actors range from individuals and protest groups to governments. After all, '[t]he Olympics also serve as a venue in which states protest the action of other states and have historically provided a means for the international community to punish or coerce states and other actors' (ibid., p. 2). Cottrell and Nelson examined qualitatively all Olympics-related protest activity that has occurred between 1896, the first modern Olympics, to the Beijing Olympics in 2008. They 'find that there has been a steady growth in protest activity since 1896, that the nature of Olympic protest has shifted from a predominance of boycotts by states and bans of states to on-site demonstrations by transnational activists, that the scope of issues covered by this protest has significantly broadened and intensified over time, and that resistance to protest by host states and the [...] IOC has been a constant factor throughout' (ibid.). The four Olympic host cities examined in these volumes have experience Olympic related protest activity that Chapter 7 dedicates some time in discussing. In order though to fully appreciate the important role that some actors give to the Games for protest activism, the following paragraphs give coverage on a selection of Olympics-related protest acts.

The following act of dissent during the first modern Olympics is mentioned by Belam (2008, p. 6) in his brief history of Olympic dissent:

> In the day following the Marathon race, a Greek woman, Stamata Revithi, ran the same course [as the male athletes] to demonstrate that it could be done by a woman. She was prevented from crossing the official finishing line inside the stadium, but is supposed to have recorded a time of around five-and-a-half hours. Women had to wait another four years for their first shot at an Olympic title.

During the 1906 intercalated Games, also hosted in Athens, Irish athlete Peter O'Connor objected to being placed on the British team. He climbed the flagpole at his medal ceremony in order to wave an Irish flag. During the 1908 London Games, the Finnish team refused to march behind the Russian flag; in addition, there was consternation because the American and Swedish flags were missing from displays around the stadium (ibid., p. 10).

The 1936 Berlin Olympics – also known as the Nazi Olympics – attracted the first calls for a boycott of the Games. There were two opposing camps in the USA, one feeling that participation in the Games showed support for Hitler's anti-Semitic policies and the other believing that going and winning would undermine Hitler's racial superiority theories (ibid., p. 15). The victory of black athlete Jessie Owens in the 100-metre sprint appeared to vindicate the latter position.

Indeed, the crowd's reaction to Owens' stunning performance showed that the German people did not have issues in exhibiting their admiration for the achievements of a non-white athlete. Indeed, it was Owens' treatment back home in the USA that left a lot to be desired. Belam writes, 'There was a ticker-tape parade in New York in his honour, but to attend the reception, as an African-American, he had to use the back elevator at the Waldorf-Astoria Hotel where it was held. The front one was reserved for white people' (ibid., p. 18).

A student prank by Barry Larking and a group of his friends, in demonstration against the torch relay around Australia during the 1956 Melbourne Olympics, gained global attention:

> A few of my friends and I thought too much was being made of this Olympic torch business. It was being treated as a god, whereas in fact it was originally invented by the Nazis for the Berlin Games in 1936. So we got a chair leg, some silver paint and an old plumpudding can, and we made our own torch.
>
> (Ibid., p. 20)

The history of protest continues. During the two decades after 1968, a symbolic year for protest politics, the 'Olympic Games were plunged into [...] crisis caused by a succession of major demonstrations, protests, boycotts, and a deadly terrorist attack' (ibid., p. 22).

Cottrell and Nelson (2010, p. 4) also argue that the Games were an important frame for the transnational activism that was marked by anti-globalization protests in Seattle in 1999:

> Several studies from the broader literature on transnational contention provide valuable insights regarding the importance of venues for protest. Some focus on how the historic backdrop of globalization has fuelled the transnational civil society and the coordinated protest against the neoliberal model at major international financial meetings.

Similarly, Lenskyj (2008, p. 2) attributes this rise in Olympic contestation to the following causes:

> The first eight years of the new millennium have witnessed ongoing campaigns by antiglobalization activists to address the

ever-growing gap between the 'haves' and the 'have nots; that is in large part a result of global capitalism. Within the antiglobalization movement are the activists in Olympic bid and host cities who continue to monitor relationships between the Olympic industry and global capitalism, most notably, the gentrification of low-income, inner city neighbourhoods and the inflation of rents and real estate prices.

The overview of Olympic contestation since the first modern Olympics of 1896 has demonstrated that as the Games were progressing to becoming the globalized event that they are today, so the issues that stimulate civil contestation in Olympic host cities moved from the local/national to the global/universal. That way it was expected that the anti-globalization movement with its extensive use of new media and campaigning in relation to 'universal world views', such as human rights and environmentalism (see Roche, 2000), would have taken advantage of the opportunities for message dissemination afforded by this global event.

The Olympic Games could not have escaped from the civil contestation that emerged with the economic downturn of 2008 and the action repertoires that have characterized responses to this crisis. Indeed, protesters have set up camps in central squares across many cities across the world, including Athens and London, thus, rather consciously or not, disputing Olympic pretensions to equality. The Olympic Tent Village that was set by the anti-Olympics movement in Vancouver was an example of a protest camp that made explicit connections between the demands of the alter-globalization movement[1] and the Olympics. In fact, Vancouver's hosting of the Games ended up rejuvenating that same protest milieu (Boykoff, 2011, p. 59). This is not surprising: 'the Olympics undoubtedly gave Vancouver activists a positive boost and refreshed their ranks with energetic younger protesters who were given a once-in-a-lifetime opportunity to soar over the hurdles that might have been present during "normal" political times' (ibid.).

Following the core theoretical perspectives of social movements theory, Cottrell and Nelson (2010, p. 4) suggest that 'the Olympics are also important because they provide a political opportunity structure (POS) (see McAdam, 1996; Tarrow, 1994) unlike any other on the world stage. Aside from their institutional component and its role in providing a POS, a crucial part on the impact of social mobilizations lies in the ways in which protest groups have mobilized the resources at their disposal. Leaving aside the usual internal sources such as money, access to available technology, expertise, and skills, "external opportunity structures" such as venues can be exploited by weak or disorganized challengers'. In agreement with perspectives that were offered by Tarrow (2005), Della Porta et al. (2006), and Della Porta (2007), Cottrell and Nelson (ibid., pp. 4–5) write that 'the choice of venue for protest can matter greatly in this regard, as a site such as the Olympics offers greater potential for domestic actors to "externalize" and enter the global arena, which can in turn lead to the formation of transnational coalitions.' Has protest and contestation ever been transformed into productive change? Sometimes contestation is channelled into new forms of social organization. Civil society actors in the form of environmental non-governmental organizations are key stakeholders in the initiation and implementation of effective strategies for environmental protection. The development of these organizations, which marks the period of the institutionalization of environmentalism, was not smooth. In fact, the rise of environmental concern took a few decades to be accepted by institutional politics after the emergence of modern environmentalism in the 1970s. Chapter 3 will provide an overview of these developments and discuss their relation to the Games and their environmental impact.

3
The Environmental Issue: Opposing Modernity and Progress

This chapter traces the emergence and development of environmental concern around the world. It starts with a brief account of the emergence of the first nature protection/conservationist associations of the nineteenth century in both Europe and the US while at the same time accounting for differences between these associations and various non-Western perspectives. It then moves to the 1960s and 70s, which can broadly be defined as a landmark period for a new extension of the environmental problematic. At this time, various social movements encouraged the politicization of the environmental issue. Thus, the chapter provides a detailed presentation of the discursive dynamics that accompanied these developments. All of these movements of the 1960s and 70s advocated limiting or even abandoning the obsession with growth that permeates all modern societies, even if the various groups identified different culprits (e.g. industrialism, capitalism, hierarchical social relations). In this context, environmentalists saw the Olympic Games and other mega-projects as extremely negative developments for the preservation of the natural environment, and they acted accordingly to prevent them from happening. It was only in the early 80s and 90s that a growing body of evidence suggested that certain advanced industrialized countries were capable, through the use of new technologies, to factor environmental consequences into both the input and output of their industrial processes without inhibiting economic growth. This period marked the beginning of the rise in environmental awareness within IOC requirements among prospective host nations.

The early antecedents of environmental concern around the world

As I said at the outset, concern for the natural environment can be traced back to ancient Greek society. The early Romans and Chinese also grappled with the environmental issue. The Rome of 20 BC had nearly 1,000,000 inhabitants and a severely polluted Tiber River. It was in this context of unparalleled environmental transformation, which had been stimulated by the growth of the Roman Empire, that the poet Horace was inspired to boast, in an epistle to his friend, that he lived in 'harmony with nature' (Bell, 2004, pp. 149–51). In the fifth century BC, the Chinese were experiencing a period of commercial and imperialist expansion, similar to that experienced by Athens after the defeat of the Persians in 480 BC. Taoist philosophers of the time expounded a sceptical view of the world and offered as an alternative to this interest in growth 'a natural conscience they believed to be free from materialist ambition [...] in things [...] "natural": in agrarian society, in the nests of tailor-birds, in moles and rivers, and in the leaves of trees' (ibid., pp. 153–4).

Despite philosophical resistance, however, empires and markets continued to expand, and concerns about the environment came only periodically into the fore, usually at the times of cultural renaissance that often precede groundbreaking socio-economic changes – for example, intellectual work surrounding the rediscovery of classical Greece during the Enlightenment subsequently fuelled the political discourses that emerged in eighteenth- and nineteenth-century revolutions. Even at these times, however, issues of the environment rarely came to the surface. As we will see in Chapter 4, socio-economic changes also influenced Marx's approach to the environment. It was only later – in the mid-nineteenth century – that action-oriented manifestations of environmental concern made their first steps with the foundation of nature protection associations and the creation of national parks.

Indeed, we often forget that the origins of environmental activism can be traced back to the mid-nineteenth century. It was at that time that the negative outcomes of industrial expansion (e.g. factory smog, polluted rivers, and poor sanitation) were becoming increasingly apparent. The fact that we have forgotten these nineteenth-century origins lies in the considerable differences between contemporary environmentalism and the ways in which environmentalism was manifested during the nineteenth century.

For that reason, it will be helpful to highlight some of the core characteristics of that early environmental activism. The characteristics of modern environmentalism (mainly composed by young, educated middle classes), as experienced in the more economically developed northern hemisphere, have led us to forget that in the past, environmentalism had many variants and was not restricted to a specific social class. For instance, the oldest conservationist organizations in the UK (1865, The British Commons Preservation Society), the US (1892, The Sierra Club), and elsewhere (France 1854, the Society for the Protection of Nature) were founded under the tutelage of wealthy individuals with strong ties to the establishment. The main perspective that guided their initiatives was that nature conservation was beneficial for the nation's health and heritage. This underlying rationale oversaw the creation of the first national parks, including Yellowstone in the US (1872) and the Royal National Park in Australia (1879), which were soon followed by others in Canada, New Zealand, and Sweden (Sutton, 2007, pp. 95–8). This first wave of environmentalism, which had a predominantly conservationist focus, continued until the end of the 1950s.

It is important to note that socio-political studies of that period (see, for instance, the influential *Civic Culture* by Almond and Verba [1963]) point out that deferential civic societies of those times exhibited high levels of trust in expert authorities. Almond and Verba (ibid., pp. 478–9) suggest that the ideal citizenry for the maintenance of liberal democracy is one that is 'active, yet passive; involved, yet not too involved; influential, yet deferential'. Any move in the opposite direction may result in an excessive questioning of elites that may lead to an undermining of democratic institutions. For Almond and Verba, British civic culture was the ideal civic culture. Political scientists who studied the newly democratized countries of the European South – Greece, Spain, and Portugal – also promoted this perception.

Today, however, the academic community is conscious of the fact that the 1950s were an exceptionally calm period in British history. The miners' strikes of the 1980s (Eatwell, 1997, p. 64) as well as the poll-tax riots and anti-motorway construction mobilizations of the 1990s clearly demonstrated that the British have acted in far different ways than the deferential citizens of the 1950s. Nevertheless, according to the theoretical perspective of reflexive modernization (RM), this belief about an older period of deference and trust in expert authorities seems to have resurfaced (see Beck, 1992, 1995; Giddens, 1990, 1991). In fact, it can be argued that this belief constitutes an essential component of reflexivity. In a different vein, Wynne (1996) has argued that, rather than accepting a past lack of contestation as evidence of trust, 'it is much more sensible to view it as indicative of an opportunity structure lacking sufficient access points by which grassroots actors might challenge the experts'. Still, the overall assessment by the two initiators of the RM perspective – that there has been a marked change in social attitudes, in the Western world since the 1960s – stands without significant controversy.

The rise of environmental concern in the 1960s and 70s

The period that followed the docile 1950s was marked by the expansion of welfare provisions and new access to higher education in Western Europe. It was a period during which more immediate needs were to a large extent satisfied, and this facilitated the opening of the path toward the higher levels of the Maslowian pyramid (Maslow, 1970). The political scientist Ronald Inglehart (1971, 1977, 1990) adopted this idea in his thesis of post-materialism. In his view, the process of 'self-actualisation' was behind the mobilization of social action around qualitative demands such as peace, the environment, and gender and racial equality. In particular, the close link between post-material values and environmentalism was first substantiated by Cotgrove (1982) in his study of new environmental groups such as Friends of the Earth, within which a commitment to environmentalism went hand-in-hand with a broader sense of alternative values.

Inglehart's post-materialism has been paramount in moving debate beyond the short-sightedness of the 'reflection hypothesis', which was adopted by some to explain the emergence of environmental concern. According to the latter perspective, the upswing in environmental consciousness after the 1970s was a direct reaction to the visible acceleration of environmental deterioration in Western industrial nations (Hannigan, 1995, p. 23). This position seems straightforward and commonsensical, but not all environmental negativities are visible to the naked eye. Furthermore, even in cases where they are visible, a whole range of other issues (e.g. employment, economic growth) can interrupt the transformation of this realization into actual social contestation. Still, attempts have been made to identify connections between the visible worsening of the environment and an increase in environmental consciousness among the public in Western democracies. The analysis of 20 years of polling data by Dunlap and Scarce (1990), for instance, suggested that a large majority of Americans believed that there was a substantial increase in threats to quality of life due to environmental deterioration. Moreover, although Jehlicka (1992; cited in Martell 1994) has argued that environmental concern is directly related to the seriousness of ecological deterioration by providing examples from different national contexts in northern and western Europe, the reality is that there is little support for the 'reflection hypothesis' in other related data. As Hannigan (1995, p. 24) has pointed out in his 'social constructionist perspective' of environmental sociology, public perception about the seriousness and extent of environmental problems 'does not necessarily reflect the reality of actual problems but rather the view of scientific experts, environmentalists and the media'. With that in mind, I want to examine some of the other perspectives that either complement or challenge Inglehart's post-materialist thesis.

Debating post-materialism and environmental concern

The idea of 'post-materialist' environmental concern has been challenged in a number of ways. Most obviously, environmental concern is not restricted to advanced industrial countries. The post-materialist thesis cannot account for the environmentalism of the poor in the global north and south. Greater amounts of wealth do not necessarily cause a decline in materialist demands or correlate with declining concern for economic and security threats. Furthermore, only a highly restricted understanding of ecological concerns can reduce these concerns to the issues of a post-materialist way of thinking.

Can we argue that concern for the environment occurs irrespective of social class? The truth is that for a long time environmentalists were drawn disproportionally from 'social and cultural specialists, individuals in creative and/or public service oriented jobs – teachers, social workers, professors etc.' (Cotgrove and Duff, 1981; Kriesi, 1989).

This finding is codified in the New Middle Class (NMC) thesis. The NMC thesis is a clear companion to the post-materialist thesis, but it places a greater emphasis on the social location of those who adopt an environmental ethic. According to one explanation, individuals placed in these occupational categories are well placed to witness victimization. Alternatively, they may have chosen their occupations for the very reason that they were already post-materialist in their value orientation. This is illustrated in the widespread involvement by Catholic priests who practise liberation theology in Latin American countries and elsewhere.

Peter Berger has challenged this apparently altruistic orientation of the NMC (1986). For him, the neo-corporatist closure from the decision-making process that these educated individuals have experienced better explains their prowess for protest politics. For Hannigan (1995, p. 27) this kind of activism in new social movements (NSMs) may even lead to 'jobs in universities, government departments, regulatory agencies and pressure groups, research grants, conference travel, etc. [and] it is not surprising that members of the new middle-class make up the bulk of the constituency of support for environmentalism, feminism, anti-nuclear activism, etc.'. Still, middle-class involvement was also common in many 'old' social movements. Since, according to the Maslowian hierarchy, environmental concern is a 'luxury', it makes sense that the middle and upper social classes will exhibit higher levels of environmental concern (Maslow, 1970, p. 183). Similarly, Van Liere and Dunlap (1980, p. 184) advance the position that the environmental concern of these classes is simply the outcome of their general concern for social problems.

The theory of 'relative deprivation' offered by Morrison and colleagues (1972) argues that the middle and upper classes share a common perspective of the 'good life' and as such are more likely to worry about the deterioration of the environment. There is also the view supported by Buttel and Flinn (1978) that complements the social class thesis by proving that, of all class-related independent variables that have an impact upon the dependent variable of environmental concern, it is the educational level that stands out. Buttell and Flinn also suggest that because the lower classes live in environmentally downgraded areas, they are the ones who should in fact exhibit higher levels of concern. This is the explanation that runs through analyses of the environmental justice movement (EJM).

The emergence of the EJM can be traced back to the US of the 1980s. It has been seen as a rejuvenation of the civil rights movement within a different contextual framework (see Capek, 1993; Pellow, 2000, 2004). Within the context of the EJM, contestation has been encouraged because of the disproportionate impacts upon communities of low socio-economic leverage on the part of Locally Unwanted/Undesirable Land Use (LULU), such as waste disposal facilities and toxic waste dumps. Nevertheless, it consistently has been proven that people react favourably to the presence of LULUs if tangible benefits can be accrued to the local community and the costs have been fairly and equitably distributed (Lober, 1993, 1995). Considering that the civil contestation has substantially contributed to increased costs for industries, businesses, and other interests that make extensive use of LULUs, the flight of these sites to developing countries with less stringent environmental and legal frameworks has increasingly become the case.

Another perspective, which has resonated with attempts to explain the manifestation of increased environmental consciousness in relation to the emergence and development of Green political formations, is a structural explanation that identifies the closure of the political system to new political actors in certain Western European democracies during the 1970s and early 1980s. In this context, NSMs are seen as defensive reactions to increased state encroachment on everyday life, a process that Habermas (1989) calls 'the colonization of the life world'. A different structural explanation pinpoints the tripartite model of interest intermediation known as 'neo-corporatism'. The participant parties in this arrangement are the state, big businesses, and big labour unions who together formulate political and economic decisions behind closed doors. Environmental groups and their concerns have been notoriously excluded from this context. More generally, the progressive adoption of the sustainable development (SD) paradigm by a range of governments across Europe since the 1990s has forced a move away from the exclusion of environmental groups from the policy decisions. This important development is intimately linked to the exploratory framework that I have proposed for assessing the capacity of Olympic Games hosting in facilitating ecological modernization.

NSMs and environmentalism

There was evident politicization and radicalization of environmental action during the late 1960s and 1970s. This action coalesced with another set of protest action frames – among them, peace and gender and racial equality. There was also significant movement overlap that allowed for the development of conceptual terms, such as 'NSM', that attempt to include a distinctive range of political/social contestation. This theoretical interest can be traced to the neo-Marxist Frankfurt School. According to Frankfurt School theorists, these movements are 'new' because they include a different set of demands to those of the 'old' working-class movements. Furthermore, their locus of action and target is 'civil society' rather than the state (see Offe, 1985; Melucci, 1980, 1985, 1988).

At the same time, protest activity is the main medium of demand articulation for NSMs (Dalton et al., 1990). Scott (1990, p. 19) suggests that, whereas the workers' movement was increasingly located within the polity through the electoral success of social democratic parties, NSMs both operate and value their location within civil society. The aims of former movement were 'political integration' and 'economic rights', while NSMs aspire to 'changes in values and lifestyle' and seek to defend civil society from statist encroaches. According to Melucci (1988, p. 248), NSM action 'takes place principally on symbolic ground, by means of a challenging and upsetting of the dominant codes upon which social relations are founded in high-density informational systems. The mere existence of a symbolic challenge is in itself a method of unmasking the dominant codes, a different way of perceiving and naming the world'. Furthermore, while the workers' movement adopted an organizational structure that was formal and hierarchical, NSMs value horizontal networks and bottom-up organizational frameworks. As such, the workers' movement focused on political mobilizations, whereas NSMs favour direct action and cultural innovation.

According to Scott, however, the view of a novelty factor in ecology, peace and women's movements is problematic. He argues that '[a]n attempt to define new movements as essentially concerned with cultural questions is to assume too unambiguous a division between questions of personal/group autonomy and more "traditional", "narrow" political issues and demands. The personal is not political merely in the sense that power relations are embedded in personal ones, but also in the sense that demands for personal autonomy, freedom, etc., are political in nature' (Scott, ibid., p. 23). It is true that attempts to challenge the newness of later movements by tracing them to earlier ones that formed around similar issues, such as the Suffragettes or the Levellers, is not particularly convincing, but taking the old worker's movement to task for not having the same approach to civil society, organizational structures, and mediums of action as that of the NSMs is ahistorical and misleading. It is equally short-sighted to assume that NSMs have been more successful than the old workers' movement in resisting the temptation of institutionaliza-tion. Indeed, students of these movements' politics have argued that NSMs do not have an exclusive orientation. They do not present a straightforward choice between a logic that aims at power and a logic that aims at identity maintenance, but rather representation of both logics with different weighting, depending on the period of action and the specific current of the movement (Rucht, 1990, p. 164).

Of course, the new political parties that are the evolutionary out-comes of the work of NSMs also appear to break the mould of tradi-tional political organization. They have characteristically been called 'non-parties' because they subscribe to 'anti-partyist' organizational structures and rhetoric, especially in the initial stages of their devel-opment. Die Grünen – the German Greens – are a paradigmatic case of a political formation that emerged out of protest politics and reac-tion toward the exclusion of NSM claims from neo-corporatist forms of interest intermediation. Because of their organizational ethos and commitment to protest politics, Die Grünen have experienced intense inner struggles over questions of organizational rationalization and their membership in larger political coalitions. It is important to note here that the well-known motto of Die Grünen, 'neither left nor right but in front', has been used within academic accounts of the phe-nomenon more that it was ever believed by its inventors. Certainly, if one actually wants to find Greens who are committed to politics that are separate from traditional divisions, one has to look away from the German case. It is paradoxical that a motto that has been used by the most leftist green party in Europe has been taken as evidence for a rupture with old political divisions. That is why the term the 'new left' has remained the most appropriate descriptor of the incorporation of the qualitative/non-materialist dimension into the left spectrum.

What is the actual relationship between environmental movements and green party formation? According to a study on the west European environmental movement:

> Green parties have both a smaller and a wider scope than the rest of the environmental movement. Their scope is smaller as they generally do not represent the environmental movement as a whole, but just a (more or less radical) part of it and wider as most parties green parties are not just environmental parties but generally constitute translation of other components of the 'new social movements sector'.
>
> (van der Heijden et al., 1992, p. 1)

Poguntke's conception of 'new politics' (1987, p. 76) provides the rationale for an explicit model of green party categorization. His view is that, without such a model, comparative research on green parties runs the risk of incorporating political formations of a conservative or centrist character, which hardly conform to the phenomenon of new politics. Poguntke suggests that, in terms of party formation, there are three possible outcomes after the emergence of the new politics conflict within a given society: the transformation of a small left-wing party into a 'new politics' party; the onset of factionalist conflict in the ranks of a larger leftist party, which may result in the breakaway of the post-materialist faction and the formation of a new political party; or the coalescing of NSM activists in the formation of a new political party. There is also a good chance that a combination of these developments may occur (ibid., p. 79). What is accepted here is that green parties are not always the sole representatives of the 'new politics'. Different national political systems and social-political contexts may produce various outcomes of party formation. Thus any examination of the trajectory of a given green party that fails to examine the relationship of the left with the 'new politics' of a specific national context is unlikely to be accurate. This is of extreme importance when we consider the translation of post-materialist concern into appeals for votes on the part of a minority of the citizenry in any given European country.

The 'new politics' formulation has been widely used in the categorization of small left and green parties. It is certainly not, however, universally accepted. Kitschelt regards the concept of 'new politics' parties' as a 'vague and undescriptive terminology, which has given rise to serious conceptual confusions and misunderstandings' (Kitschelt and Hellemans, 1990, p. 473). Instead, he talks about left-libertarian political parties (LLPs). Kitschelt (1988, p. 197) offers the following definition of LLPs:

> Consistent with the socialist legacy, left-libertarians are 'left'; they oppose the market place and insist on solidarity and equality. They are also 'libertarian' in that they reject centralized bureaucracies and call for individual autonomy, participation, and the self-governance of decentralized communities.

Kitschelt wishes to place unequivocally those political phenomena on the left of the political spectrum. It may be the case that the 'new politics' transcend old polarities between left and right, as well as that the meaning of the left has itself become multidimensional. A cultural dimension that emphasizes qualitative issues such as individual autonomy, decentralization and environmental protection has also enriched the left's conception of itself. Indeed, the concept of left-libertarian politics encapsulates this important differentiation.

There is no doubt that there is great use in devising explicit formulations, but an analysis of both theses – the idea of a 'new politics' and one that is 'left-libertarian' – reveals that there are no substantial differences between them. In fact, Müller-Rommel (1990, p. 211) points out that the 'new politics' label is 'largely synonymous with the left-libertarian parties that Kitschelt discusses'. Indeed, Kitschelt's (1990, pp. 186–7) criteria for the separation of left-libertarian parties from traditional political forces are virtually identical to those put forward by Poguntke and Müller-Rommel. In addition, the organizational ethic of left-libertarian parties rejects the hierarchy and formalization that characterize mass political parties.

The new politics/LLP family can certainly be viewed as what Katz and Mair (1995, pp. 23–4) call 'challengers' to the cartel party, the latter being a manifestation of an 'ever closer symbiosis' (ibid., p. 6) between parties and the state and a successor to the classic catch-all party formulation put forward by Kirchheimer (1966). After all, the exclusion both of new politics and LLP constituencies from systems of interest intermediation as well as their anti-partyist organizational structures and rhetoric certainly exclude them from the cartel party family. However, Katz and Mair (ibid., p. 24) are careful to point out that, even in the case 'of many of the Green parties, where the opposition is more deep-rooted ... demands also prove capable of accommodation and cooption'. Indeed, the participation of green parties in government coalitions during the 1990s can be perceived as indicative of a transformation of their rhetoric along more compromising lines. Nevertheless, we know that this transformation did not take place without opposition from the more fundamentalist elements in those parties, albeit with different degrees of intensity depending on the national context.

Ultimately, the behaviour of a political party depends upon its primary goal. According to Strøm (1990, pp. 570–1), there are three types of competitive party behaviour: vote-seeking, office-seeking and policy-seeking. This chapter has already highlighted the anti-partyist logic that characterizes green parties at the initial stages of their formation. As such, they belong to that category of political parties that claim policy/ideological and intra-party democracy maximization as their primary goals (Harmel and Janda, 1994, pp. 269–71). For Strøm (1990, p. 568), the policy-seeking party is the 'least adequately developed model of competitive party behaviour'. In that case, office-seeking becomes a means to an end (that is, influencing policy) rather than a goal *per se*. Thus, intra-party competition in the ranks of these parties arises over the extent to which office acquisition justifies changes to the overall strategy of the party, including its insistence on intra-party democracy maximization.

I have demonstrated how the emergence and rise of environmental concern made an impact upon political mobilization. Attention was also paid to the choices made by protest actors and their progressive entry into institutional politics. That account was necessary in order to illustrate the passage of environmental concern into the corridors of power in certain advanced industrialized nations. These nations have become the pacesetters for environmental policies across Europe. This was important in laying the foundations for exploring various interrelated developments in the following chapters, such as the inclusion of the environmental factor in IOC guidelines for Olympic Game hosting, the general development of environmental sociology in parallel with the development of different manifestations of environmental concern across recent decades, and the analytical dynamic that is offered by reflexivity in investigating the potential for ecological modernization within Olympic host nations.

Cross-national determinants of environmental concern

Concern for the environment can take many shapes and forms, ranging from simple expressions of interest to behavioural changes that are aimed at environmental protection. Of course, the degree of importance of such behavioural changes can be determined from their outcomes as much as from their objectives (see Stern, 2000).

The general label of 'environmental deterioration', which is often used in order to group together a variety of environmental problems, disregards the fact that not all of these problems are of equal importance. At the same time, the notion of 'environmental interest' or 'environmental concern' lacks specificity and appears not to take into consideration the fact that there are many different ways to express such interest, some of which carry substantial costs. Clearly, different kinds of 'environmental concern' are implied by the following statements: (a) 'I am worried about GM food stuff'; (b) 'I am worried about GM food stuff and only buy organic products'; and (c) 'I am worried about GM food stuff and that's why I am destroying GM crops'. Additionally, it has been shown that levels of concern for different types of environmental deterioration exhibit significant variation and that demographic variables (e.g. age, gender) give different weight to different types of environmental deterioration (Van Liere and Dunlap, 1981, pp. 663–7).

Another important methodological issue is how to measure environmental concern as well as the extent to which we can draw cross-national generalizations about such data. Undoubtedly, the objective reality of environmental deterioration (e.g. visible pollution of a local river) is going to influence local environmental concerns; thus, any comparison with other regions is likely to be problematic. Recent research (Marquart-Pyatt, 2007) has demonstrated that the existence of differences both within and between countries depends intimately upon which expressions of concern one attempts to study. For instance, environmental concern in France was the highest among 15 countries surveyed when the question was connected to social institutions, but it fell to eleventh place when environmental concern was set against prices and employment. These marked variations have played an important role in assessing the environmental concern that has been exhibited by the populations of Olympic host nations in the context of the austerity measures that governments have adopted in response to the recent global economic crisis.

Limits to Growth advocacy

I have so far said very little about the ideas that inspired the development of environmental awareness. Public concern about the protection of the natural environment and the rapid diminishment of key sources for the maintenance of energy inputs can be traced to the late 1960s and early 1970s. In this context, the *Limits to Growth* debate of the 1970s very much shaped the environmental concern and activism that were to follow.

The *Limits to Growth* report (1972) was produced by a team of scientists led by Donatella Meadows and commissioned by the Club of Rome. Using computer modelling, the team explored a number of different scenarios (12 in total) for different global trends, such as the acceleration of industrialization, population growth, widespread malnutrition, the deterioration of natural environments, and the depletion of non-renewable resources. In concluding their report, Meadows' team painted an apocalyptic picture about the capacity of societies to maintain their growth dynamics without doing anything about rising environmental crises. In particular, they claimed that 'if nothing was done, then even if the amount of available resources were doubled, pollution was reduced to pre-1970s levels and new technologies were introduced into the model, economic growth would still end before 2100. Some commentators saw this as a vindication of the radical ecological position that industrial societies were just not sustainable' (Sutton, 2007, p. 133). Characteristically, Dobson (1995, p. 35) has argued that the *Limits to Growth* report is an undisputed 'symbol for the birth of ecologism in its fully contemporary guise'. Or as Eckersley claimed (1992, p. 8), 'the notion that there might be ecological limits to economic growth that could not be overcome by human, technological ingenuity and better planning was not seriously entertained' until after the publication of the report'. Although various perspectives pointing to the negative aspects of continuous growth without care for ecological limits have re-appeared, this time the concern about climate change is accompanied with proposals for technological fixes that can mitigate climate change rather than a complete overhaul of the existing system.

Blueprint for survival

During the same period, another publication added to this pessimistic story about the environmental costs of economic growth. The *Blueprint for Survival* (Goldsmith et al., 1972) was published as a special issue of *The Ecologist*. The authors of the *Blueprint* stated: 'our Blueprint of survival heralds the formation for the movement for survival and, it is hoped, the dawn of a new age in which Man will learn to live with the rest of nature rather than against it' (ibid., p. 10). As Spaargaren (2000, pp. 41–2) later claimed, 'The Blueprint contained not only a warning for the future of mankind but also a model of an alternative, green society. A society consisting of numerous small scale units, where people live their lives close to nature and to each other, where technology was of the proper scale.' Put more simply, the *Blueprint* authors argued that '[t]he principal defect of the industrial way of life with its ethos of expansion is that it is not sustainable' (Goldsmith et al., 1972, p. 15).

It is interesting to note that the *Blueprint* was adopted at the same time as the manifesto of the UK Green Party. This radical and apocalyptic political position was marked by a negation of the old political structure in both its left and right formulations. The fact that this discourse, which had intense spiritualist undertones, played a pivotal role in the UK exemplifies the differences in key ideological components that have marked the emergence and development of green parties across Europe. Evidently, the well-known German Green Party slogan, 'neither left nor right but in front', was in many respects closer to the British case than to other green party formations.

There are still other 'streams of thought which underpinned the strategies and ideologies of the environmental movement from the early seventies onwards' (Spaargaren, 2000, p. 42). A number of authors known collectively as 'theorists of counter-productivity' (e.g., Ivan Illich, Rudolf Bahro, André Gorz, and Barry Commoner among others) have been influential within the environmental movement and have envisioned alternative societies. The following paragraphs describe the nature of this alternative society in the work of two influential authors, E. Fritz Schumacher and Murray Bookchin.

Schumacher

In *Small is Beautiful* (1973), Schumacher calls for a return to smaller-scale, self-sufficient communities. In that book, Schumacher formulates the following identification:

> The illusion of unlimited powers, nourished by astonishing scientific and technological achievements, has produced the concurrent illusion of having solved the problem of production. The latter illusion is based on the failure to distinguish between income and capital where this distinction matters most. Every economist and businessman is familiar with the distinction, and applies it conscientiously and with considerable subtlety to all economic affairs – except where it really matters: namely, the irreplaceable capital which man has not made, but simply found, and without which he can do nothing.
>
> (Schumacher, 1988 [1973], p. 11)

In relation to this claim, Schumacher suggests the need to move away from 'the technology of mass production' to 'the technology of production by the masses, making use of the best modern knowledge and experience, [which] is conducive to decentralisation, compatible with the laws of ecology, gentle in its use of scarce resources, and designed to serve the human person instead of making him the servant of machines' (ibid., p. 128).

These claims have been very popular in the early stage of green party formation where the 'anti-partyist' perspective coupled with a belief in the need for radical societal change dominated the green discussions. In general, *Small is Beautiful*, is viewed by many as the work behind their interest in environmental issues and its legacy survives through the activities of the Schumacher Society, Schumacher College and Intermediate Technology Development Group (ITDG) with many achievements in the developing world (Spowers, (2002, p. 287).

Bookchin

Murray Bookchin (1982, 1990, 1991) formulated the idea of Social Ecology alongside his re-discovery of Athenian democracy and a connection between the ecological imperative of 'unity in diversity' and anarchist thinking. Bookchin offered a way of organizing the autonomous, self-sufficient communities that had been envisaged by Schumacher. He proposes a new, stateless social order that he calls 'libertarian municipalism'. Such an order is 'a confederal society based on the co-ordination of municipalities in a bottom-up system of administration as distinguished from the top-down rule of the nation-state' (as cited in Barry, 2007, p. 172). Bookchin goes on to describe the characteristics of this society as follows:

> Property [...] would be shared and, in the best of circumstances, belong to the community as a whole, not to producers ('workers') or owners ('capitalists'). In an ecological society composed of a 'Commune of communes', property would belong, ultimately, neither to private producers nor to a nation-state. The Soviet Union gave rise to an overbearing bureaucracy; the anarcho-syndicalist vision to competing 'worker-controlled' factories that ultimately had to be knitted together by a labor bureaucracy. From the standpoint of social ecology, property 'interests' would become generalized, not reconstituted in different conflicting or unmanageable forms. They would be *municipalized*, rather than nationalized or privatized. Workers, farmers, professionals, and the like would thus deal with *municipalized* property as citizens, not as members of a vocational or social group.
>
> (Bookchin, 1993, p. 373)

The organizational model proposed by Bookchin had been influential in the most radical sectors of the environmental movement and green party factions. Together with Schumacher, the small-scale organization and immediate-democratic decision-making advocated by Bookchin still echoes in the gatherings and writings of alter-globalization activists.

Overall, the views of all of these authors (classified together as the 'limits to growth' theorists) 'have been very influential within the environmental movement' (Spaargaren, 2000, p. 42). Nevertheless, by the early 1980s, '[t]here were good reasons for the environmental movement to reconsider its relationship with the state, not only on theoretical but also on empirical grounds. During the seventies and the beginning of the eighties, the environment not only settled itself on the top of the political agenda, one could also witness the substantial growth of the body of environmental legislation' (ibid., p. 44). According to Mol (2000, p. 48), there has been a substantial change in the guiding ideologies of ENGOs 'from, the mid 1980s onward, particularly in the industrialised countries of Western Europe'. In particular, as I pointed out earlier on, the countercultural perspectives envisaging a radical break with industrialization that had influenced the early environmental made way to radical reformism. 'Although fierce opposition against the capitalist economic system, against industrialization and large or complex technosystems, and against any form of large bureaucracy can still be found in the diverse ideological spectrum of the environmental movement, these ideas have definitely moved from a core position to the periphery between the 1970s and the 1990s' (ibid.).

Changes of the 1980s and early 1990s

Beginning in the early 1980s, there has been an increase in environmental legislation as well as a *de facto* incorporation of environmental themes into the politics of certain advanced western European countries. This new legislation and ideological incorporation is both the outcome of and a contributor to the institutionalization of the environmental movement. As Hajer (1996, p. 252) writes:

> Typically the political conflict is also seen as a learning process. 'We owe the greens something', it is argued. The dyed-in-the-wool radicals of the 1970s had a point but failed to get it through. This was partly due to the rather unqualified nature of their *Totalkritik*. The new consensus on ecological modernization is here attributed to a process of maturation of the environmental movement: after a radical phase the issue was taken off the streets and the movement became institutionalized as so many social movements before it. With the adoption of the discourse of ecological modernization its protagonists now speak the proper language and have been integrated in the advisory boards where they fulfil a 'tremendously important' role showing how we can design new international forms to come to terms with environmental problems. Likewise, the new consensus around ecological modernisation has made it possible that the arguments of individual scientists that found themselves shouting in the dark during the 1970s are now channelled into the policymaking process.

These developments have led to the establishment of EM as a dominant theory of environmental sociology (Buttel, 2000a, 2000b; Mol, 2000). Thus, the following two chapters offer a detailed examination of sociological theory in general and its relationship to the environmental factor. Aside from describing a historical change of attitude among environmentalist actors toward a more moderate campaign of reform and accommodation within existing institutions, this chapter serves as a prelude to the story about collaboration of some ENGOs with the IOC with the goal of conducting green Olympic Games.

4
Environmental Concern and Environmental Sociology: Parallel Developments

This chapter identifies the extensive parallels that exist between the development of environmental concern and the emergence and development of environmental sociology. The first cogent critique of indefinite economic growth that does not conserve or replenish the natural resources that provide its essential inputs arose during the 1970s. This critique came as a continuation of earlier warnings that various industrial techniques and processes were having a deleterious impact on the natural environment. The result was a convergence of two currents of environmental concern: the first brought to attention the limits of developmental processes themselves and the second pointed out their adverse impact on natural resources with no immediate exchange value. The contestations of various environmental movements at the time were marked by powerful negations of modernity in general and of the capitalist mode of production in particular.

At the same time that these political concerns were taking shape, environmental sociology made its first steps towards establishing itself as a sub-discipline of sociology. This new sub-discipline promoted the incorporation of the biological sciences into the larger discipline's analysis of nature-society relations. According to the founders of environmental sociology, Catton and Dunlap, ordinary sociology was guided by the view that the role of sociologists consisted in the exploration of social facts and that biological and ecological laws had nothing to contribute to their analysis. As a result, Catton and Dunlap proposed a shift towards a New Ecological Paradigm (NEP) that recognized the ecosystem dependency of all human societies and

51

a re-examination of the fundamental assumptions that underpinned the discipline of sociology.

It was not long before a number of social scientists proclaimed that classical sociological theory, as bequeathed to us by Marx and Durkheim, itself offered significant ecological insights, which for various reasons had been overlooked in the past (John Bellamy Foster), or that classical theory provided an extremely useful framework for the social examination of the current environmental crisis (Raymond Murphy, Peter Dickens, Albert Schnaiberg). These perspectives are still current, but they have been confronted by newer perspectives that have achieved considerable attention and placed the environmental problematique at the epicentre of current theory. I am referring to the reflexive modernization (RM) perspective that was first formulated by the German sociologist Ulrich Beck. This perspective was much prefigured by developments on the environmental front that took place in a number of advanced nations, and it caused a serious blow to earlier perspectives that had suggested there was a basic incompatibility between environmental preservation and the modernity project. In what follows, I provide a brief history of the parallel developments of environmental awareness and environmental sociology.

The emergence of environmental sociology[1]

The birth of environmental sociology is intimately linked to an increase in environmental awareness in the Western world. This new awareness was exemplified by a number of key events of the late 1960s and early 1970s. For example, the years 1972 and 1973 were marked by the key publications of the Club of Rome, *The Limits to Growth*, and the *Blueprint for Survival*, as well as the foundation of Friends of the Earth. Only a year later, this activity was followed by the foundation of what is today the most popular environmental protest group on Earth, Greenpeace. In addition, the decade began with the most populous public protest gatherings that had occurred to that time, namely Earth Day in the USA, which included the participation of 300,000 demonstrators. Moreover, in 1972 the UN Conference on the Human Environment brought out into the open the main challenges that faced any global agreement about environmental protection, namely the disparity between developmental needs and this protection. It was at that conference that a Brazilian delegate famously said that he 'prayed for the day when they would share in the developed world's industrial pollution and would welcome multinational investors, willing to help them pollute' (Leonard, 1988, p. 69). That statement played a paramount role in generating the basis for the sustainable development (SD) principle during the early 1990s. It also confirmed a more important point: that the standard sociological examination of socio-structural differentiation was more than relevant in the study of environmental issues.

It was in this context that the first hesitant steps of the new sub-discipline were made. Publications by Catton and Dunlap (see Catton and Dunlap, 1978 and Dunlap and Catton, 1979) and Buttel (1978) argued for the need to bring biophysical variables into the sociological endeavour through the employment of what Catton and Dunlap called the 'new ecological paradigm' (NEP) and Buttel 'new human ecology' (NHE). I want to take a minute to describe each of these theories and their importance for the new environmental sociology.

Buttel (2002, p. 36) has argued that most histories of environmental sociology began 'almost by ritual or reflex, with a critical commentary on the past century or so of sociology and how it had gone wrong in dispatching the biophysical environmental as a set of phenomena worthy of sociological interest'. Indeed, the way in 'which sociology came to define itself, especially in relation to potentially competing disciplines such as biology and psychology, effectively excluded or forced to the margins of the discipline ... questions about the relations between society and its "natural" or "material" substrate' (Benton, 1994, p. 29). This exclusion of the natural from the endeavour of sociology was also true of what Goldblatt (1996, p. 2) sees as the 'second wave of classical social theory'. According to Goldblatt, however, for the founding fathers of sociology – Auguste Comte and Herbert Spencer – sociology was not considered independent of biology but was highly dependent, or even subordinate, to it. In fact, both Comte and Spencer modelled their explanations of social development on biological and Darwinian insights. In effect, it was in the third quarter of the nineteenth century that the exclusion of the natural from the social was put into action. Developments at the time had made redundant the more dismal predictions of the discipline, such as the near apocalyptic predictions of population growth made by Malthus and the limits to growth alarm bells that were rung by Ricardo and Stuart Mill. Goldblatt (ibid. p. 3) summarizes this situation:

> In any case, a more pressing issue for sociology was always going to be its supposed dependence on or subordination to the natural sciences in general and biology in particular. If sociology was to emerge as a distinctive body of knowledge, then its subject matter – society – had to be cordoned off from the realm of biology and nature.

This perspective and its influence on the relationship of the discipline with the field of the natural have been showcased in apt detail in Raymond Murphy's (1995) 'Sociology as if nature did not matter'. There, Murphy identifies a 'pre-ecological sociology' that was active at a time when environmental concern and awareness were only beginning to attain prominence. Murphy identified a set of internal and external factors that explain the absence of nature from the discipline of sociology. Among internal factors, Murphy castigated the specialization of the sociological discipline. Sociologists, he argued, had to establish sociology as the preeminent discipline for the examination of the social. In their effort to do so, they disregarded environmental parameters:

> It is not that sociologists have tried to take the dynamics of nature into account in their theories but were prevented from doing so by obstacles to interdisciplinary research; rather the vast majority have not tried. In order to give added importance to social action, sociologists have assumed that the effect of nature to social action can be ignored. They have obscured the context of social action, thereby de-naturing humans, their society and culture. Sociologists assuming the social construction of reality have the same epistemological status as carpenters assuming the world is made of wood.
>
> (Ibid., pp. 695–6)

A second internal factor identified is what Murphy calls *revulsion and repulsion*. Revulsion and repulsion are, in a nutshell, the reaction of the sociological community to any attempt to explain the social by means of the natural. That reaction led, in Murphy's claims, to the further distancing of the discipline from the investigation of social action in conjunction with natural processes. In particular, this revulsion and repulsion targeted the perspectives promoted in the work of the sociologists that Goldblatt (1996) calls the 'first wave of social theory'. Herbert Spencer, for instance, was inspired by the environmental factor (deriving from his understanding of Darwin's theory of evolution) to develop an ultra-conservative social theory.

A third internal factor, according to Murphy, was *an aggravating tendency* in the discipline to neglect the relationship between natural and social processes. The main reason behind this neglect is the relatively recent distancing of sociological theory from the investigation of social systems and the study of 'meaningful interpretation (including notions of such systems)' (ibid., p. 699).

Murphy's last factor, this time external, is the social and economic framework in which sociology emerged and developed as a scientific discipline. This was a period when the negative outcomes of the manipulation of nature had not yet become evident. The overexploitation of the natural world on behalf of societal advances was to continue without interruption. This was the context in which sociology was initially constructed.

The premises of environmental sociology

Catton, Dunlap, and to a certain extent Buttel have all attempted to overcome these restrictions to sociological engagement with the natural. Catton and Dunlap suggested that any attempt to bypass these restrictions would have to come into confrontation with the whole gamut of principal sociological theories, since all of them were conditioned by the mentality of unhinged growth that had then permeated the world's wealthiest nations.

The New Ecological Paradigm (NEP)

Catton and Dunlap published a number of articles during the late 1970s in an effort to found this new sub-discipline. Their proposal envisaged a radical restructuring of the discipline. They separated sociological engagement with environmental problems into two categories. The first dealt with issues ranging from nature recreation and management of environmental resources to the emergence of environmental concern and the environmental movement. The second recognized 'the fact that physical environments can influence (and in turn be influenced by) human societies and behavior' (Dunlap and Catton, 1979, p. 244). Dunlap (2002, pp. 12–13) later elaborated upon the authors' motivation for making this distinction:

> First, I wanted to show that some sociologists were going beyond treating environmental problems as socio-political issues and were, in fact, employing environmental variables to examine societal-environmental relations, which I felt was essential for demarcating environmental sociology as a distinct area of specialization akin to political sociology, for example. Second, I was fearful that if sociologists limited themselves to investigating environmental attitudes and activism and the processes by which environmental quality came to be defined as a social problem, our nascent field would possibly atrophy when societal attention to environmental problems declined.

The latter point, on the possible atrophication of environmental sociology due to its dependency on issue-attention cycles (Downs, 1972), has continuously come to the fore in periods of intense fluctuations in public environmental concern and decline like that experienced with the publication of the Intergovernmental Panel on Climate Change (IPCC) report on climate change in 2006 and the advent of the financial crisis in 2008. However, as it's going to be demonstrated in Chapter 5, the prominent role played by the environmental issue in reflexivity has elevated environmental sociology to a permanent position in the sociological endeavour.

Similarly, according to Buttel (1987, p. 466), the main objective of environmental sociology was to redirect research by means of a sociology that 'recognized the role of physical-biological factors in shaping social structures and behaviors, [a sociology] that was aware of the impacts of social organization and social change in the natural environment. Environmental sociologists sought nothing less than the reorientation of sociology toward a more holistic perspective that would conceptualize social processes within the context of the biosphere'.

Catton and Dunlap's first article (1978) claimed that the various theoretical viewpoints within the discipline tended to overemphasize their conceptual differences when, in point of fact, these differences were inconsequential when compared to a common commitment to anthropocentrism (ibid., p. 42). All of these theories, according to Catton and Dunlap, fall under the Human Exemptionalism Paradigm (HEP). The critique of HEP targets the perception of human exemptionalism. The authors understand HEP as the result of 'the ensemble of values and ideologies that have predominated during the 500-plus years' boom of Western expansion underwritten by finite supplies of fossil fuels and non-renewable raw materials' (Buttel, 1987, p. 469). The growing awareness of environmental problems in the 1970s brought to the fore the realization that 'human societies necessarily exploit surrounding ecosystems in order to survive, but societies that flourish to the extent of overexploiting the ecosystem may destroy the basis of their own survival' (Dunlap and Catton, 1979, p. 250). This is the perception that underpins the NEP, a model whose premises highlight the ecosystem dependency of all human societies (Catton and Dunlap, 1978, p. 45).

There are three basic premises to the NEP:

1. *Homo sapiens* is just one species among many, all of which are weaved and entangled in the biotic communities that construct our social lives.
2. Complicated connections between cause and effect as well as feedback to and from the natural 'complex' result in many unintended consequences to intentional human actions.
3. The world is finite; thus, there are strong natural and biological limits that constrain economic development, social progress and other social phenomena.

To put it simply, under the prism of the NEP, the environmental sociology proposed by Dunlap and Catton examine the relationship between the physical environment and the social complex.

The analytical framework proposed for this study is that of the 'ecological complex'. The ecological complex forms an analogy with the biological concept of an ecosystem. As such, the complex can be understood through the acronym POET: it consists of the interdependence between the population (P), social organization (O), including cultural, social, and personality systems, the natural environment (E), and technology (T). The POET concept was originally created by Duncan (1961). Its roots lie in orthodox human ecology, and it can be seen as an 'embryonic form of environmental sociology' (Buttel and Humphrey, as cited in Hannigan, 1995, p. 16).

Both the NEP and POET examine the mutually reinforcing relationship between the environment and human beings. The three functions performed by that environment – as supply depot, as living space and as waste repository – can be overexploited. Their capacity for replenishment and renewal can be affected to such an extent that life for humans and other animals, as well as for local, national, and global ecosystems, becomes precarious. In order to illustrate how these functions overlap, we can bring to mind how the overuse of one natural resource by the supply depot function can have an impact on the other two functions. Shortages in drinking water resulting from the overuse of aquifers, for instance, may result in mass population migrations to other regions. This migration can then result in a host of other problems as it is bound to impinge significantly upon the living space or habitat of those communities (in terms of housing and transportation systems, for example). This in turn results in an increase in waste by-products that have the potential to exceed the capacity of that ecosystem to function as a waste repository. This framework is ecological because it assumes an impact in one area or function has impacts on other areas or functions with which it is in contact.

Catton and Dunlap's perception that the entire gamut of sociological theory was incapable of producing a worthy analysis of environmental problems was strongly challenged by Buttel (1978) in his call for a new human ecology (NHE). He later argued (1986, as cited in Buttel, 1987, p. 469) that

> the major classical sociological theorists were concerned with natural and biological phenomena to a far greater degree than is typically acknowledged by environmental and nonenvironmental sociologists. That the contributions of the classical theorists to what is now referred to as environmental sociology have remained unrecognized can be largely explained by the fact that most sociologists have limited acquaintance with the full range of the classical theorists' primary writings and have learned the classical tradition through secondary treatments by sociologists with little or no interest in environmental and biological phenomena.

After years of continuous in-fighting about these issues, Dunlap (2002a, p. 19) made the following clarification:

> We were not suggesting that existing theoretical perspectives were no longer relevant in general, nor even that they were irrelevant for analyzing environmental issues [...]. Rather, we were trying to emphasize that unless our discipline managed to shed the blinders imposed by the HEP, we feared that environmental issues would not have been taken seriously by sociologists.

All in all, NEP was, according to Catton and Dunlap, an attempt 'to stimulate development of more ecologically sensitive or greener versions [of existing theoretical perspectives]. Clearly, then, [They] were implying that NEP-oriented theories will not be totally incommensurable with older theories, but simply be grounded in more realistic assumptions about the relationship between modern societies and the biophysical environment' (Dunlap, 2002b, p. 343).

Resurrecting the classics

Catton and Dunlap produced a truly powerful perspective that critically confronted the undisputed dominance of anthropocentrism across the social sciences. Nevertheless, it did not take long for the scholarly community to identify ways that classical sociology might still be relevant in contributing to discussion of environmental problems. This conversation produced three tendencies:

1. One group of sociologists was sceptical of the position, widespread in environmental sociology circles, that dismissed the tradition of the classic triumvirate (Durkheim, Marx, Weber) and its potential relation to the treatment of the environmental. This group was also partly behind the NEP advocacy of Catton and Dunlap. As such, these sociologists consciously worked toward extracting 'ecological' insights from the work of classical thinkers (Catton; Foster).
2. Another group has embarked upon the selection and application of concepts 'from the collected works of the sociological pioneers' (social class, status, alienation etc.) in their analytical treatment of the current environmental crisis (Dickens; Murphy).

3. We have now witnessed a typological approach whereby the field is organized on the basis of classical theory. Hannigan (2006, p. 6) uses the example of Sunderlin (2003), who 'defines and conceptualises three key paradigms (individualist, managerial, class) each of which is derived from the classical sociological literature (Durkheim, Weber, Marx)'.

Durkheim

The accusation that the classical sociological tradition is inadequate for investigation of the environmental issue has been mainly directed toward the Durkheimian social facts *dictum*. Durkheim appeared to have made substantial inroads into questions of ecology through his proposal of a societal bifurcation between 'mechanical' and 'organic solidarity'. The latter idea, as Catton has suggested (2002b, p. 90), is 'fundamentally ecological, based on the *interdependence* of a society's members resulting from the diversity of various members' roles in achieving collective adaptation to the circumstances in which life is lived'. Despite this apparently proto-ecological concept, however, Durkheim's effort to establish sociology as a scientific discipline was accompanied by an insistence on the 'objective reality of social facts' and the importance of explaining social phenomena by means of the exclusive use of social variables. Given this background, it is not surprising that the earliest proponents of environmental sociology declared that environmental sociologists share 'a common interest in the physical environment as a factor that may influence (or be influenced by) social behavior. Thus, researchers focusing on the physical environment, whether built or natural, share in a mutual departure from the Durkheimian dictum that social facts must be explained with other social facts' (Dunlap and Catton, 1979, p. 255). This was, in essence, an outcome of Durkheim's decision to 'elevate *social facts* over "facts of lower order"' (that is, psychological or biological facts) (Hannigan, 2006, p. 6).

Still, efforts have been made to rehabilitate Durkheim's place among those who think about environmental issues. We have already mentioned his evolutionary model for societies in *The Division of Labour in Society* (1893) that distinguishes between states of *mechanical solidarity* and *organic solidarity*. As Gross (2000, p. 480) points out,

Durkheim based his explanation of differentiation in society on the factor of an assured natural increase in population and problems arising from this fact due to a limited amount of space and natural resources. This is obviously an environmental (humans depend on natural resources) rather than a pure sociological type of explanation.

(Gross, ibid.)

Indeed, Gross goes on to remark that, in *Rules of Sociological Method* (1895), 'Durkheim's definition of society includes the material environment, that is, roads, railways, houses, waterways, clothes, and as he generally termed it, *things as a matter of course*' (ibid., pp. 280–1). The social environment, which constitutes the source of social processes, 'is composed by elements "of two kinds: things and persons" and social things "are of the same nature as physiological phenomena"' (ibid., p. 281). Although Durkheim did not hesitate to '[deny] these material things the original motivating power for social transformations', he did later argue that 'there is a need to take them into account in the explanations which we attempt' (ibid.).

Durkheim's proto-environmentalist position on these issues should not, however, be overstated. Although, for Durkheim, society is considered to be part of nature, it occupies a higher position than the rest of nature. And although he made partial inroads in his first works, where he stressed the need to take into consideration the possible impact that the environment might have on the composition of society, he failed to further develop this line of thought. In his later works, changes to the environment that have been caused by societal actions ceased to have negative connotations. All in all, it is clear that Durkheim abandoned his earlier suggestion that the natural environment was capable of having an impact upon the construction of society. In Durkheim's view, nature simply has a regular and static character (see Botetzagias, 2008b, pp. 32–3).

This view can partly be understood by thinking through the historical context in which Durkheim was working. As Goldblatt (1996, p. 4) puts it,

> the mid-nineteenth to early twentieth centuries appear as an ambiguous moment in the ecological history of modern societies. Durkheim explained the emergence of modernity and its unique economic and social capacities in terms of a progressive division of labour and political stratification, in which individuals and societies could find their competitive niche; population and resource

pressures spurred the technological innovation and social differentiation on which modern industrial society was built. Once stratified it was the abnormal forms of the division of labour that upset the balance of modern societies rather than natural resource constraints.

According to Catton (2002, p. 93), Durkheim's theorization of the nature-society relationship was affected by selective reading of Darwin, and the unavailability of any knowledge of ecology and evolution in Darwin's time. As Hannigan (2006, p. 7) puts it, '[w]hat we are left with then is chiefly speculation on what *might have been*. [Some, like Järvikowski (1996, p. 82) do not hesitate to argue that] Durkheim would likely have written in a different way today about the relations between the social and physical environment because biological theory has undergone a profound process of change'.

Weber

We are now moving to an examination on how the environment fared in the theoretical endeavour of the second father of sociology. Weber is of particular importance in this study due to his key role as an inspiring source behind the formulation of reflexivity that is used extensively as an analytical framework here.

According to Goldblatt (1996, p. 3),

[o]f the classical trinity, Weber's work conducts the most limited engagement with the natural world. There are some reflections on the environmental origins and implications of nomadism in his study of Judaism. Yet his historical investigations of antiquity, despite the centrality of agrarian production in his work, yielded little direct study of the historical impact and social implications of differing natural environments.

Still, other researchers have not hesitated to suggest that, '[u]nlike his contemporary Durkheim, Weber had no reluctance to admit the causal significance of non-social factors for social processes' (Albrow, as cited in Murphy, 1995, p. 694). Or as West writes (cited in Murphy, 2002, p. 74), 'Weber's ecological analysis emphasised the interactive role of geography, climate, natural resources, and the material aspects of technology in the structure and change of historical social structures'. In the exploration of Weber's environmental contribution that is performed below, extensive use has been made of work produced by Raymond Murphy, who has developed a neo-Weberian environmental sociology. It is important to highlight from the beginning that, although Weber emphasized the role of values and agency,

> he did not propose a reductionism to the social. He held that culture was grounded in, even if not determined by, nature and to take the social out of the realm of natural causality altogether was to confuse the ideal and dogmatic formulations of jurists [and we might add sociologists] with empirical reality.
>
> (Albrow, as cited in Murphy, 1995, p. 257)

This sentiment has compelled Murphy (2002, p. 75) to argue that '[u]nderstanding the distinctive element of intentions among humans does not require the neglect of nature and the effect of its ecosystems on social action'. Thus, 'it is not correct to view Weber simply as an "interpretivist" sociologist, at least if this implies that he neglected the material bases of social life' (Sutton, 2004, p. 4).

Weberian scholarship has challenged such an interpretation of Weber's work and highlighted the need to acknowledge his 'ecological materialism' (ibid., p. 5). Indeed, the instrumental rationality, composed by calculability, efficiency and the employment of science, if left inflexible in their appreciation of the natural 'can be closely associated with ecological irrationality (Murphy, 2002, p. 79).

Marx

Last but not least, Marx offers a perspective that could have been a cornerstone in the sociological engagement with the environment should it have been brought out into the fore by sociologists at a much earlier stage. The following paragraphs offer a detailed exploration.

'For the early Marx the only nature relevant to the understanding of history is human nature ... Marx left nature (other than human nature) alone.' So said George Lichtheim, in his influential *Marxism: An Historical and Critical Study* (1964). With these words by

Lichtheim, John Bellamy Foster starts his account on Marx and the natural. Foster writes that

> [t]hough he was not a Marxist, Lichtheim's view here did not differ from the general outlook of Western Marxism at the time he was writing. Yet this same outlook would be regarded by most socialists today as laughable. After decades of explorations on Marx's contributions to ecological discussions and publications of his scientific-technical notebooks, it is no longer a question of whether Marx addressed nature, and so did throughout his life, but whether he can be said to have developed an understanding of the nature-society dialectic that constitutes a crucial starting point for understanding the ecological crisis of capitalist society.
>
> (Foster, 2002)

Similarly, Goldblatt (1996, p. 4) has argued that 'Marx had acquired a persistent concern with the notion of the natural from his study of Hegel … [and] placed the economic interface of human societies and the natural world at the centre of historical change'. Moreover, he continues, 'Engels, as early as the 1840s, found the urban environment to be aesthetically repugnant and an active contributor to the misery of the poor. Marx himself was aware of the capacity of capitalism to undermine soil fertility and abuse natural resources. But these reflections have only derived significance in retrospect' (ibid., p. 5).

This quote partly substantiates the point made at the start of this section about the missed chance to bring the environmental factor to the epicentre of the sociological endeavour. This issue was eloquently brought out into the open by Foster.

Indeed in a seminal article on the issue that he produced earlier, Foster argues that:

> Marx provided a powerful analysis of the main ecological crisis of his day – the problem of soil fertility within capitalist agriculture – as well as commenting on other major ecological crises of his time (the loss of forests, the pollution of the cities, and the Malthusian spectre of overpopulation). In doing so, he raised fundamental issues about the antagonism of town and country, the necessity of

ecological sustainability, and what he called the 'metabolic' relation
between human beings and nature.

(Foster, 1999, p. 373)

For Marx,

> Capitalist production collects the population together in great
> centres, and causes the urban population to achieve an ever-
> growing preponderance. This has two results. On the one hand it
> concentrates the historical motive force of society; on the other
> hand, it disturbs the metabolic interaction between man and the
> earth, i.e. it prevents the return to the soil of its constituent ele-
> ments consumed by man in the form of food and clothing; hence
> it hinders the operation of the eternal natural condition for the
> lasting fertility of the soil ... All progress in the capitalist agricul-
> ture is a progress in the art, not only of robbing the worker, but of
> robbing the soil; all progress in increasing the fertility of the soil
> for a given time is a progress toward ruining the more long-lasting
> sources of that fertility ... Capitalist production, therefore, only
> develops the techniques and the degree of combination of the
> social process of production by simultaneously undermining the
> original sources of all wealth – the soil and the worker.

(Marx, 1976 as cited in Foster, 1999, p. 379)

For Dickens (2004, p. 80), 'Marx's early background led him to
undertake no less than an analysis of what would now be called
"environmental sustainability". In particular, he developed the idea
of a "rift" in the metabolic relation between humanity and nature,
once seen as an emergent feature of capitalist society'. With this argument Dickens connects Marx's concerns to contemporary debates
about the environment. In addition, Dickens directs our attention
to another aspect of a metabolic rift that connects Marx to issues of
mega-events and the environment. Inspired by Davis (2002), he puts
forward the following arguments that might be connected to the
imperatives of cities hosting the Olympics Games:

> Three metabolic problems have developed in the modern city.
> These are the provisions of an adequate water supply, the effective

disposal of sewage and the control of air pollution, problems not unknown before the present time ... This is another instance, therefore, of humanity's metabolism with nature not being ultimately destroyed but being overloaded in the context of a particular kind of social and spatial organization.

(Dickens, 2004, pp. 84–85)

As we would see in due course, all three of these metabolic problems are taken into account in the bids made by prospective Olympic host cities.

But can this metabolic rift be addressed through the modernization of the production process? This is definitely something that reflexivity perspectives support (a topic that will be discussed in Chapter 5). Indeed, the examination of Olympic host nations is a kind of testing ground for this hypothesis.

Marxist approaches to the environmental problem

Readings of Marx's understanding of the relationship between the natural and social have been sources of tension for the ecological movement. The former communist bloc had a catastrophic environmental impact (see, for example, Dickens, 2004), and the overall perception in Western Marxism was not that dissimilar from the Promethean attitude that dominated the interpretation of Marxism in the Soviet Union (Foster, 2002a).

Some ecologically minded scholars have, however, made good use of Marx. They made extensive efforts to change the perception that Marxist theory is hostile to the natural. To be sure, environmental issues were not a primary concern in Marx's work, and nobody can accuse Marx of failing to predict the ecological crisis of the twenty-first century.

With this in mind, we can pass into an examination of neo-Marxist approaches to the society-nature relationship.

Metabolic rift (John Bellamy Foster)

According to Buttel et al. (2002, p. 7), '[t]he prominence of neo-Marxism in environmental sociology can also be gauged by the fact that arguably the most prominent environmental-sociological journal article in the history of the field thus far has been that by Foster (1999) in the *American Journal of Sociology*, in which he advances Marx's notion of "metabolic rift" as a promising focal point for the development of environmental sociology'. In that article, Foster (1999, p. 399) claims that '[t]he way in which Marx's analysis prefigured some of the most advanced ecological analysis of the late 20th century – particularly in relation to issues of the soil and the ecology of cities – is nothing less than startling'. In the current phase of globalized capitalist flows, this process of an ecological rift is even more acute, and it has had an immense impact on environmental sustainability (e.g. depletion of the ozone layer, groundwater pollution and deforestation). Yet, it is in the context of these global capitalist flows that the International Olympic Committee (IOC) has operationalized the green dynamic imbued by Olympic Games hosting. Is there no hope for a post-Olympic green legacy in a capitalist system? This is an issue that to a certain extent this book puts to the test. Meanwhile, I will discuss the fellow-travelling perspectives put by two other Marxist scholars in order to further substantiate the Marxist political economy critique that we are confronted with at a later stage.

The second contradiction of capitalism (James O'Connor)

According to James O'Connor, capitalism suffers from an innate tendency to undermine the very conditions that sustain its existence, and this occurs in a way that was not recognized by Marx. According to Marx, the first contradiction of capitalism involves the increased mechanization of production, which leads to the eventual deskilling of the labour force and, consequently, layoffs. Because workers are also consumers and their ability to consume is curtailed by the deskilling process, capitalists are then unable to realize surplus value or profit. Growing discontent leads to the organization of the labour movement and the eventual transition to a communist society. Unlike Marx, O'Connor claims that this crisis of capitalism can also occur because of the increased ecological costs that result from its unstoppable exploitation of natural resources:

> Global capitalist development since WWII would have been impossible without deforestation, air and water pollution, pollution of the atmosphere, global warming and the other ecological

disasters; without the construction of megacities, with no regard for congestion, rational land use and transport systems, and housing and rents; and finally, without the reckless disregard for community and family health, physical and emotional, education and other 'components' of the socialized reproduction of labor power – not to speak of the welfare of future generations.

(O'Connor, 1998, p. 10)

For these reasons, capitalism is bound to reach a point whereby the overexploitation of the natural resources that are essential for its reproduction – the second contradiction of capitalism – create barriers to profit realization, rampant social contestation and an eventual transition to red-green socialism. A very similar perspective to that of O'Connor has been formulated by Alan Schnaiberg.

The treadmill of production (Alan Schnaiberg)

For Schnaiberg (1980), environmental damage cannot be reduced simply to the inevitable outcome of industrialized processes. In fact, the inherent grow-or-perish dynamic of capitalism is a more appropriate culprit. This is the 'treadmill of production':

Competitive pressures lead to a constant motivation for more and cheaper production of goods and inevitably, perhaps, a lack of concern for environmental protection, which would be more costly and prevent companies from being able to compete effectively. That means a lack of incentives within the economic system for environmental protection and a green light for pollution, waste and environmental damage, as long as this enhances profitability.

(Sutton, 2007, p. 67)

Along with the metabolic rift, the treadmill of production perspective involves an important spatial dimension, 'which leads to the constant conflicts over development that are familiar to local communities everywhere' (Bell, 2004, p. 57). Indeed, Molotch (1976, p. 310) has argued that

the desire for growth provides the key operative motivation toward consensus for members of politically mobilized local elites, however split they might be on other issues, and that a common interest in growth is the overriding commonality among important people in a given locale – at least insofar as they have any important local goals at all. Further, this growth imperative is the most important constraint upon available options for local initiative in social and economic reform. It is thus that I argue that the very essence of a locality is its operation as a growth machine.

Following Logan and Molotch (1987), Bell (ibid.) argues that growth machines are 'dedicated to encouraging almost any kind of economic development – frequently with little regard for environmental consequences or the wishes of affected neighbourhoods. Local people have spatially fixed investments of a different sort, and because of them conflict with growth coalitions often arises'.

Concluding remarks

This chapter discussed the emergence and development of environmental sociology. I identified a close parallel between the historical rise of environmental concern and the first skirmishes of this new sub-discipline. I continued by arguing that the founding fathers of environmental sociology (Catton and Dunlap, and Buttel) held the conviction that the established discipline, as influenced by Durkheim, Weber and Marx, had short-changed environmental issues. Despite this defining claim, a careful re-reading of these classical theorists substantiated that they *had* taken into account the fact of the biological. In fact, for certain theorists, the biological metaphor was so essential that its absence would have brought down their whole theoretical edifices. Durkheim's organic solidarity analogy is illustrative. His social facts' injunction has given the impression that he was hostile to the biological dimensions of sociality, but research shows that this is far from the case. It is Weber who demonstrated a rather limited engagement with biology and the potential impacts on society that may be occasioned by changes to the natural environment. In the end, however, Karl Marx appears to have had the most groundbreaking impact on the sub-discipline through the adoption an eco-Marxist perspective. This perspective has been used as a counterweight to the positions and projections of the EM outlook, which is the subject of the next chapter.

5
Reflexive Modernization: Connecting the Environment with Modernity and Modernization

This chapter offers an analytical overview of Reflexive Modernization (RM). It provides a detailed presentation of the two constituent dimensions of RM: the risk society and the ecological modernization (EM) perspectives. It discusses how both dimensions resonate with certain approaches to environmentalism and, more particularly, how EM sits alongside the discourses that have been used to promote the environmental credentials of sport mega-events such as the Olympic Games.

Reflexive modernization

The concept of RM claims that we are confronted with a new phase in social development whereby scientific knowledge needs to be used to overcome the negative outcomes of earlier forms of technological intervention. RM entered the sociological lexicon through the writings of Ulrich Beck's *Risk Society* (1992; 1994; 1995; 1999). It also resonated with the work of Anthony Giddens (1990; 1991). RM is essentially Beck's understanding of the challenges with which the world is confronted in late modernity. This situation is characterized by a disintegration of frameworks that were previously taken for granted, including family and community networks, effective employment and trust in expert authorities. This disintegration has resulted in a perpetual struggle over definitions, within which the subpolitical plays a paramount role and challenges the established policy frameworks of liberal democracies (Buttel, 2000a, p. 29).

Throughout the 1990s, the second dimension of RM was fleshed out in the work of a group of environmental sociologists at Wageningen

University in the Netherlands. This work became known as Ecological Modernization (EM), and was influenced by the work of two German social scientists, the sociologist Joseph Huber and the political scientist Martin Jänicke.

This chapter argues that the two dimensions of RM – Risk Society and Ecological Modernization (EM) – share two important premises:

1. Individual and institutional choices are not simply reflections of the dynamics that result from the dominant structures of capitalism and industrialization.
2. The resolution of environmental problems will not emanate from the 'de-modernization' or 'anti-modernization' that is advocated by the most radical environmentalists but from a gradual modernization of all societies.

(Buttel, 2000a, p. 29)

Notwithstanding this convergence, however, there are points of significant divergence between the two dimensions. Although neither has room for the claims made by radical environmentalists, the idea of a risk society takes a more positive view of the role of new social movements (NSMs). By the same token, although EM tends to view the state as being capable of implementing certain environmental improvements, Beck's risk society is much more sceptical of state intervention (Buttel, 2000b, p. 36).

Still, the two models share a similar expectation about the relationship of modernity and ecology (Blowers, 1997; Cohen, 1997). Although Beck's model is openly critical of key aspects of modernity, it cannot be seen as a dystopian exhortation. In fact, it could be argued that Beck's reflexivity is a call to the further modernization of modernity. Similarly, EM advocates that the negative outcomes of modernity are the result of the fact that the modernizing project has not been left to take its 'natural' course.

Risk society

The umbrella concept of the risk society encompasses a phase in the development of modern societies in which the social, political, ecological, and individual risks created by the momentum of innovation increasingly elude the control and protective institutions of those societies (Beck, 1999, p. 72). In his formulation of this concept, Beck employs a grand theoretical narrative made up of three different historical epochs:

1. 'pre-industrial', or traditional society, which is characterized by mainly so-called 'natural hazards' (droughts, floods etc.);
2. industrial society', or first modernity, within which natural disasters are complemented by constantly increasing man-made dangers (e.g. smoking, drinking, occupational injury) but within which exist advancements in scientific knowledge as well as the ability to regulate both natural disasters and man-made risks;
3. 'risk society', a second or advanced modernity, within which the dominant risks of advanced modernity (e.g. air pollution, chemical warfare, biotechnology) stem from industrial or techno-scientific activities.

(see Mythen, 2004, p. 18–19)

Finally, although Beck makes extensive use of the concepts of risk and danger in relation to many facets of social life, his engagement with these concepts revolves around the general issue of environmental deterioration. As Goldblatt (1996, p. 155) has written, 'one could go as far as to argue that the risk society is predicated on and defined by the emergence of these distinctively new and distinctively problematic hazards'.

The importance of these ideas for environmental sociology cannot be overstated. For Hannigan (1995, p. 12), Beck's theoretical contribution has the potential to 'lift environmental sociology into the mainstream of debate into the broader field of sociology', whilst Irwin (2001, p. 51) suggests that

Beck's exploration of the 'risk society' raises [...] a fundamental set of problems for both science and political institutions in dealing with environmental issues. His account also suggests a profound role for the 'social' and for sociological investigation within what he portrays as the current environmental crisis.

Although both facets of RM propose the need for the further modernization of modernity, the concept of the risk society appears to offer a more radical perspective in its contribution to social analysis. As Beck (1992, p. 52) himself writes,

the problems emerging here cannot be mastered by increased production, redistribution or expansion of social protection – as in the nineteenth century – but instead require either a focused and massive 'policy of counter-interpretation' or a fundamental rethinking and reprogramming of the prevailing paradigm of modernization.

Societies have always been threatened by dangers and hazards. Following Giddens (1991), however, we can say that what conditions these threats as *risks* is that their dangers are known, and their materialization can be both predicted and calculated. Moreover, the production and the consequences of risks are both qualitatively and quantitatively different from earlier forms of risk. Industrial societies of the past were confronted by both spatially- and socially-confined risks. The contemporary forms of environmental degradation that Beck focuses on 'are not spatially limited in the range of their impact or socially confined to particular communities' (Goldblatt, 1996, p. 158). The idea that risks are equalized in contemporary societies (exemplified by Beck's famous maxim that 'poverty is hierarchical, smog is democratic' [1992, p. 36]) has been the subject of much criticism, but it is a powerful idea with repercussions for the study of the Olympics and the environment. Before talking a bit about this, however, I will continue to explore the placement of risks across time in Beck's work.

From industrial society to risk society

Beck determines the occurrence of dangers and threats across time in order to provide support for his narrative of social development. This narrative begins with the pre-industrial (or traditional) society, continues with the industrial society (or first modernity), and culminates with the risk society (a second, late modernity). In order to illustrate the differences among these periods, Beck employs exemplary cases of risks. For instance, he argues that, in pre-industrial societies, risk took the form of natural disasters (e.g. earthquakes, droughts).

Responsibility for these catastrophes did not lie with a single guilty party, and these catastrophes were neither seen as accidents nor as the outcome of negligence but simply as inevitable dangers. Their origin was mainly attributed to external, even supernatural forces (e.g. God, demons, nature). Calls for the moderation of or escape from their consequences were always directed toward these forces (Beck, 1995, p. 78; Goldblatt, 1996, p. 159). By contrast, in the industrial society, the comprehension of and the proposed methods for confronting risks are directly linked to individual action as well as to general societal forces. At this stage of development, societal forces have the ability to control and restrict the consequences of both natural catastrophes and technical risks (e.g. smoking, alcohol consumption, occupational injuries). This understanding is evident in the development of health and welfare systems, environmental agencies, and insurance companies (Beck, 1992, p. 98). Finally, with the transition to the risk society, there is an expansion of environmental dangers (e.g. pollution of the atmosphere, chemical warfare, biotechnology), and these risks are understood to have resulted from the techno-scientific activities that dominate social and cultural experiences (Mythen, 2004, p. 16).

In contrast to natural hazards, manufactured risks are created by human action. Indeed, because manufactured risks originate in the developmental stage of the modernization process, they can be interpreted as social rather than natural constructions. On top of the increasing complexity of modern environmental risks, societies and individuals in the later stages of modernity must cope with a wave of personal risks and uncertainties. The idea of RM connects these two processes: natural hazards and personal risks.

Science in the risk society

In his challenging formulation of the developmental paths of modernity, Beck is intensely sceptical about scientific reasoning and its role in environmental destruction. His work often functions as a full-blown indictment of scientific endeavour. He writes:

> The origin of risk consciousness in highly industrialized civilization is truly not a page of honor in the history of (natural) scientists. It came into being against a continuing barrage of scientific

denial, and it is still suppressed by it. To this day the majority of the scientists sympathize with the other side. Science has *become the projector of a global contamination of people and nature.*

(Beck, 1992, p. 70)

No doubt, the development of scientific endeavour, especially through its manifestation in applied technologies, is responsible for many of the environmental risks that confront modern societies, yet this is not the full picture. The truth is that any identification of environmental risks and their impact must still be formulated in scientific terms. Science has the capacity to offer possible solutions to these problems:

> Certainly, it is difficult to engage with the contemporary treatments of environmental issues without encountering scientific analysis and argumentation. Given the difficulties discussed by Beck of identifying and measuring environmental problems (since we cannot necessarily touch, hear, see, taste or smell them and since, in any case, they may extend beyond the boundaries of not only our senses but also of previous human experience), science has become a crucial means of apprehending these potential problems and of thinking the unthinkable.
>
> (Irwin, 2001, p. 71)

Although Beck seems to appreciate the role played by science and its 'sensory organs', he asserts that the relationship between environmental protection and science is marked by ambiguity. For him, in order for science to be placed in a position of prominence as a key contributor to the identification and provision of solutions to environmental problems, it must acknowledge the criticisms that confront it, instead of thinking of itself as superior to citizens' concerns about the environment (Beck, 1995). Yet Beck also believes that such a development is impossible because the training and research work of natural scientists is similar to the ideological indoctrination that was received by Stalinist cadres in the Soviet Union (ibid., p. 127).

> Scientific calculations of risks remain locked in a circle of technological mastery. The abstractness of the calculation in relation to particular technologies guarantees that technologies can be

compared; comparison with accepted technologies in turn guarantees calculability, which must for that very reason be unverifiably, circularly presupposed, compelling the denial of what can never be excluded: dogmatism. The assumption of technological mastery turns into technological irrationality, in the face of hazards which cannot be technically excluded but only minimized.

(Ibid., p. 126)

This position is in many ways opposed to that of EM and sustainable development (SD), within which science is seen as a force of rationalization and progress that can actually provide the means by which developmental processes can be maintained with minimal harm to the environment. In order to truly understand Beck's position on the ethical failures of science and technology, it is worth thinking about his response to the Chernobyl 'accident'.

Nuclear energy

Beck repeatedly refers to the existence of 'inconceivable risks', among which he includes the negative health and environmental impacts that are the result of the use of nuclear energy. The risks entailed in the use of nuclear energy have long been a source of public concern, largely because of a number of nuclear accidents at nuclear energy reactors (Windscale, 1957, UK; Three Mile Island, 1979, USA). However, it was the 1986 disaster in Chernobyl, which resulted in 56 deaths (more in the long term), that played a prominent part in Beck's theorization.

For a start, since no other nuclear accident had, at that time, taken such a devastating toll as Chernobyl, we can say that one of the main causes behind the accident was the institutional structure of the Soviet regime. In particular, it is better to think about the case of Chernobyl as a paradigmatic case of the irrationality that permeates the chain of command within authoritarian regimes than to understand it as an example of the unavoidability of nuclear disasters. In the Chernobyl accident, there was a severe lack of communication and trust among the various participants who were part of the operation of the reactor. Both engineers and party-appointed directors at the reactor were disinclined to report errors and/or the technical limitations of the plant because such reports would have been followed by serious disciplinary sanctions.

Beck overlooks this perspective. He has continued to make use of the Chernobyl case as a means of illustrating the uncertainty that characterizes the technological applications within late modernity. According to Beck, if applied science can only go as far as probability recording in thinking through the dangers that may result from various modernizing projects, risk and uncertainty are inseparable from techno-scientific applications. The worst-case scenario is always present. As such, all of the scientific experts who are behind the planning and operation of a nuclear reactor can only guarantee *probable* safety. Even if there were to be three "accidents" in three different nuclear reactors tomorrow, the validity of the statements made by scientific experts would remain undisturbed (Beck, 1999, p. 59–60).

The uncertainty that permeates debate about nuclear energy is evident in that debate's bipolarity. One side views nuclear energy as the dangerous face of all modern technology; the other sees nuclear energy as a relatively safe means of energy production (Irwin, 2001, p. 76). This difference of opinions on the nuclear issue appears in James Lovelock's *The Revenge of Gaia* (2006). Lovelock, the formulator of the Gaia Hypothesis, argued in his earlier work that

> [i]t is now generally accepted that man's industrial activities are fouling the nest and pose a threat to the total life of the planet which grows more ominous every year. Here, however, I part company with conventional thought. It may be that the white-hot rash of our technology will in the end prove destructive and painful for our own species, but the evidence for accepting that industrial activities either at their present level or in the future may endanger the life of Gaia as a whole, is very weak indeed.
>
> (Lovelock, 1979, p. 107)

His perspective is not much changed. In *The Revenge of Gaia*, he writes:

> I believe that nuclear power is the only source of energy that will satisfy our demands and yet not be a hazard to Gaia and interfere with its capacity to sustain a comfortable climate and atmospheric composition. This is mainly because nuclear reactions

are millions of times more energetic than chemical reactions. [...] That means that the amounts of nuclear fuel needed to supply our energy demands are tiny compared with Gaia's normal mass transactions, and so is the quantity of waste produced. We could use nuclear fission or fusion for quite some time before we ran into the kind of problem we are now having with fossil fuel.

(Lovelock, 2006, pp. 67–8)

This pro-nuclear view has a broad appeal among certain policy actors, in particular among those who advocate nuclear energy as one way of reducing CO_2 emissions and adhering to the requirements of the Kyoto Protocol. The 2007 Intergovernmental Panel on Climate Change (IPCC) report further added to this debate. It was accompanied by a stream of proposals for the development of technological adjustments, such as the use of renewable energy sources like solar and wind, biomass for heating, biofuels for transport needs, carbon capture and storage (CCS), and nuclear energy. Each of these energy sources attracted its own debate about advantages and disadvantages at the local, national and global levels. Although, nuclear energy was one of many such options and was accompanied by a range of risks (waste, disposal, security), it appeared to be experiencing a renaissance.

Still, it was unexpected to see this position endorsed by prominent British environmentalist and social commentator, George Monbiot. Most surprising was that Monbiot's endorsement was expressed at a time during which the meltdown at Fukushima Daiichi plant had reignited concerns about the safety of nuclear reactors (see Cockburn, 2011). Monbiot wrote:

You will not be surprised to hear that the events in Japan have changed my view of nuclear power. You will be surprised to hear how they have changed it. As a result of the disaster of Fukushima, I am no longer nuclear-neutral. I now support the technology. A crappy old plant with inadequate safety features was hit by a monster earthquake and vast tsunami. The electricity supply failed, knocking out the cooling system. The reactors began to explode and melt down. The disaster exposed a familiar legacy of poor design and corner-cutting. Yet, as far as I know, no one has yet received a lethal dose of radiation.

(Monbiot, 2011b)

Bemused by this position, Cockburn (2011, p. 77) exuberantly exclaimed, 'Does Monbiot live on a Fantasy Island?' Monbiot rationalized his views about nuclear power in this way:

> I despise and fear the nuclear industry as much as any other green: all experience has shown that, in most countries, the companies running it are a corner-cutting bunch of scumbags, whose business originated as a by-product of nuclear weapons manufacture. But, sound as the roots of the anti-nuclear movement are, we cannot allow historical sentiments to shield us from the bigger picture. Even when nuclear power plants go horribly wrong, they do less damage to the planet than coal-burning stations, operating normally [...]. The Chernobyl meltdown was hideous and traumatic. The official death toll so far appears to be 43: 28 workers in the initial few months then a further 15 civilians by 2005.
>
> (Monbiot, 2011a)

Cockburn (ibid., p. 77) argues that '[t]he 1986 explosion in the fourth reactor at the Chernobyl power station in the Ukraine does indeed remain the benchmark catastrophe amid peacetime nuclear disasters'. He goes on to say that the '[d]enial that Chernobyl actually killed – and is killing – hundreds of thousands of people is crucial to the efforts of the nuclear lobby. Amid the Fukushima crisis, Fergus Walsh, the BBC's medical correspondent, comforted his audience with the absurdity that by 2006, Chernobyl had prompted only sixty deaths from cancer; the same drivel has been repeated many times over since the Fukushima catastrophe [...]'(ibid.).

This disagreement between Monbiot and Cockburn gives added credence to Beck's understanding of the uncertainty that permeates contemporary environmental risks. At a time when the memory of Chernobyl appeared to have been confined to the dustbin of history, the Fukushima disaster has reignited opposition to the proposition of clean, safe nuclear energy. In Germany and France, countries with nuclear energy capacities, there have been widespread mobilizations against atomic energy. As a result of the accident at Fukushima, the planned phasing out of nuclear energy in Germany by 2022 was confirmed, and 8 of Germany's 17 nuclear reactors were shut down immediately (Gersmann, 2011). This development raises questions about Germany's ability to meet its objective of reducing greenhouse gases

(GHG) emissions by 40 per cent as of 2020, as well as concerns about its standing as a pacesetter for environmental progress in the EU.

Ecological Modernization

Ecological Modernization (EM) is the second facet of RM. Blowers (1997, p. 846) provides the helpful definition that EM 'holds that while environmental constraints must be taken fully into account, they can be accommodated by changes in production processes and institutional adaptation'. Spaargaren (2000, p. 41) has shown that 'changing theoretical perspectives go hand in hand with changing views on the relationship between environment and economic growth, on the role of science and technology and on the role of both governmental and non-governmental actors'. Therefore, in order to understand the factors that have contributed to the emergence, development and establishment of EM as a core theoretical contribution to the sociological analysis of environmental issues, we must describe the economic, political and scientific changes that helped to cause an overall change in thinking about the environment.

According to Buttel (2000b, p. 60), during the 1980s, environmental sociology was characterized by a division between a North American perspective, which viewed the environmental deterioration caused by the developmental process of the capitalist system as a given, and a new, European perspective, which took into consideration the environmental reforms that had taken place in certain northern European countries. The former had been influenced by environmental justice mobilizations against the location of hazardous facilities in disadvantaged communities in the USA. The latter was nothing other than the EM perspective. By the mid-1990s, EM had become accepted by the international environmental sociology community as a significant theoretical addition to the perspectives that had previously dominated 'eco-social' thinking, such as the 'treadmill of production' and other ideas derived from Marxist political economy.

From *Limits to Growth* to sustainable development

During the early 1980s, there was an evident increase in environmental legislation in certain European countries, especially West Germany and the Netherlands, and this legislation was coupled by the placement of environmental issues at the top of those countries' policy agendas. At the same time, there was evidence of discontinuity in the traditional practice whereby intensified economic development is interwoven with intensified environmental perturbation. A process known as the 'delinking of material flows from economic flows' began to occur. In certain cases, whether local or national, environmental reform played a crucial role in the absolute reduction of polluting emissions and the overuse of natural resources (Mol, 2000). According to van der Heijden (1999, p. 202), it is not a coincidence that the origins of EM can be traced to some of the political work of these countries (in particular, Germany and Holland). These countries experienced strong environmental movements that had a significant procedural impact as well as gaining access to the centres of power. Indeed, the populations of these countries continue to exhibit high levels of environmental consciousness.

As I elaborated upon earlier, the partial entry of the German Green party into the German parliamentary system in the early 1980s led to conflict among its factions. After years of infighting, as well as selective collaboration with the Social Democratic Party (SPD) at the state-level, they managed to form a coalition government in 1998. This is a paradigmatic case where a movement party or 'anti-party' party that employed a radical discourse heavily influenced by protest politics ended up joining with one of its opponents, the central government, in order to implement a number of institutional changes. To paraphrase Helmut Wiesenthal (1993), who played a leading part in the internal consultations of the Green party by pushing for the adoption of a realistic political strategy, the average citizen, when confronted with a millennial discourse that invokes the certainty of an imminent and non-reversible catastrophe, is like a patient who is informed about the non-reversibility of her disease and chooses to simply 'enjoy' the last days of her life. That is, confronted with the possibility of radical catastrophe, citizens may well decide to leave things as they stand. Wiesenthal's point is that, instead of espousing a radical discourse that envisions powerful ecological change when the right conditions have materialized, it is more productive to propose feasible and incremental changes. After all, such changes are a more accessible discourse to people who cannot risk siding with the revolutionary discourse and praxis that is advocated by green activists.

Joseph Huber analysed EM as a historical phase of contemporary society. According to his proposed scheme, industrial society develops in three phases that Hannigan (1995, p. 183) presents as follows: '(1) the industrial breakthrough; (2) the construction of industrial society; and (3) the ecological switchover of the industrial system through the process of "superindustrialization"'. Overall, any understanding of EM advocacy needs to highlight the fact that EM focuses on the modernization of modernity through the correction of the flawed structural design of modernity – what Mol (1996, p. 305) has called 'the institutionalized destruction of nature'. According to Hajer (1995, p. 32),

> EM is basically a modernist and technocratic approach to the environment that suggests that there is a techno-institutional fix for present problems. Indeed EM is based on many of the same institutional principles that were already discussed in the early 1970s: efficiency, technological innovation, techno-scientific management. It is also obvious that EM [...] does not address the systemic features of capitalism that make the system inherently wasteful and unmanageable.

In another work, Hajer (1996, p. 250–1) has proposed three ideal-typical interpretations of EM. These consist of the institutional, technocratic and cultural interpretations.

At the centre of Hajer's institutional interpretation is the presumption that institutional bodies can learn and adequately adjust their practices when confronted by convincing arguments that 'correspond with the scientific evidence available' (ibid., p. 252). The idea is summed up by the famous phrase, the 'hardware can be kept but the software should be changed' (ibid.). This is the sentiment that has guided assessment of the EM capacity of the Olympic host nation-states under study here.

For supporters of the institutional interpretation, modern capitalist enterprise can adjust to environmental restrictions without missing out on positive productive conditions, new markets, or even increased profit margins under conditions of control by both the state and civil society. One example of this is that environmental protection and reform have been proven to create profitable markets for the so-called green industry. In this paradigm of development, the incorporation of nature as a third force (after capital and labour) into the production process of capitalist economy has become a feasible theoretical reality (Buttel, 2000a, p. 30; Mol, 1995, p. 41). Similarly, Spaargaren (2000, p. 52–6) views EM as the emancipation of ecological rationality and the ecological sphere – that is, the independence of this sphere from its economic counterpart. This is an important first step toward the consecutive and equal incorporation of the other two sectors, the ecologization of the economy and the economization of ecology (Mol, 1996, p. 313–4; Mol and Spaargaren, 1993, p. 437).

As far as the technocratic interpretation is concerned, Hajer (1996) brings in the initial radical critique of technocracy on the part of radical environmentalists, and for the cultural interpretation he follows the anthropologist Mary Douglas in suggesting that discussions about environmental pollution must be understood as discussions about citizens' preferences for ideal types of social organization. Still, for Buttel (2000b, p. 58–59), Hajer has not managed to establish EM as a true progenitor of accord between industrial and ecological processes. He has instead offered a description of the hegemonic discourses that have permeated the environmental policies of certain advanced countries.

Similarly, Christoff (1996, p. 485) suggests that

> [i]t is possible to illuminate problems and issues left unaddressed or unresolved by [certain] uses of EM by asking a series of interrelated questions. In different situations (different policy forums and different countries), quite different styles of EM may prevail – ones which can be judged normatively to tend toward either weak or strong outcomes on a range of issues, such as ecological protection and democratic participation. In this sense, these questions hint at the limitations of those forms of ecological modernization which tend toward the first rather than the second of what might seem, initially, opposing poles.

An intimate connection between EM and SD has often been detected in sites where a weak version of EM has taken hold. In fact, this relationship has led to the treatment of the two perspectives as identical.

This is because 'ideologically and practically, such ecological modernization may simply put a green gloss on industrial development in much the same way that the term "sustainable development" has been co-opted – to suggest that industrial activity and resource use should be allowed as long as environmental side-effects are minimised' (Christoff, ibid., p. 486). This is not necessarily a cause for significant upset since there are also weak or 'hollowed-out' (Hayes and Horne, 2011, p. 751) forms of SD. Weak interpretations of SD correspond with weak interpretations of EM, not least because these interpretations comfortably coexist in international organizations such as the EU. Seen this way, it is not exaggerating to say that EM is *the* dominant discourse in political debates about environmental issues. Still, the normative and prescriptive claims of EM constitute an attractive framework for sociological analysis.

The concept of SD made its entry into international debate about the environment with the publication of *Our Common Future* by the World Committee for Environment and Development (WCED 1987) under the direction of the former Prime Minister of Norway, Gro Harlem Brundtland. The book was an attempt to combine economy, development and the environment in a single discourse that would allay the fears of developing nations that the proliferation of environmental awareness in the developed world would arrest their own developmental processes (Benton, 2002). The sentiments expressed by the delegation of Brazil at the 1972 UN Conference on the Human Environment that 'pollution was a sign of progress and that environmentalism was a luxury only developed countries could afford' (Hogan, 2000, p. 2) are well known. Like EM, SD also does not see the developmental process as inherently anti-environment. In fact, according to the well-known definition of SD, 'sustainable development is development that meets the needs of the present without compromising the ability of future generations to meet their own needs'. As is evident here, SD is characterized by its vagueness. It is therefore a very attractive concept for various forces, including governments in both developing and developed countries and multinational corporations. According to Huber (2000, p. 269),

> The NGOs understanding of sustainable development has been formulated by themselves as an anti-industrial and anti-modernist

strategy of 'sufficiency', meaning self-limitation of material needs, combined with 'industrial disarmament', withdrawal from the free world-market economy, and an egalitarian distribution of the remaining scarce resources. Contrary to that, the industry's understanding of sustainable development is the 'efficiency revolution'. Industry and business are looking for a strategy that would allow for further economic growth and ecological adaptation of industrial production at the same time. The means for achieving this goal is seen as the introduction of environmental management systems aimed at improving the environmental performance, i.e. improving the efficient use of material and energy, thus increasing resource productivity in addition to labour and capital productivity.

The environmental non-governmental organizations' (ENGOs') interpretation of SD appears to be in tune with 'the strongest and most radically *ecological* notion of EM [as that often stands] in opposition to industrial modernity's predominantly instrumental relationship to nature as exploitable resource' (Christoff, ibid. p. 494). As indicated earlier, however, governments' and corporations' interpretation of SD is more in tune with a weak EM. Indeed, the general acceptance of Local Agenda 21 (LA21) at the 1992 Rio de Janeiro UN Conference on Environment and Development, which is often seen as marking the official institutional acceptance of SD, has confirmed that a weak EM perspective dominates political discussions on environmental issues. We now see increasing collaboration between ENGOs and multinational corporations, although there are still environmental organizations that refuse to follow this path.

As a way of beginning to discuss the nature of the indicators that I have identified and selected for my examination of the potential facilitation of EM in Olympic host nations and cities, I want to discuss the five-cluster synopsis of EM themes as assembled by Mol and Sonnenfeld (2000, p. 6–7). According to Mol and Sonnenfeld, EM acknowledges:

1. the changing role of science and technology;
2. the increasing importance of market dynamics;
3. a transformation in the role of the nation-state;

4. modifications in the position, role and ideology of social movements;
5. changing discursive practices.

As in Beck's work, EM recognizes the culpability of scientific endeavour behind the emergence and proliferation of environmental problems. However, it significantly diverts from Beck's position by not only placing more emphasis on the identification and cure of environmental ills, but by attempting to prevent these ills from occurring in the first place by means of the application of innovative technological preventatives. Mol (2000, p. 46) explains this position:

> First, science and technology are not only judged to be involved in the emergence of environmental problems but they are also valued for their actual and potential role in curing and preventing them; second, traditional curative and repair options are replaced by more preventative socio-technological approaches that incorporate environmental considerations at the design stage of technological and organizational innovations; finally, the growing uncertainty of scientific and expert knowledge on definitions, causes and solutions with respect to environmental problems does not result in a marginalization of science and technology in environmental reform.

Interestingly, from the EM perspective the market can be an important facilitator of an ecological restructuring of production. This is because it is very much conditioned by the interdependent relationships of producers, customers, consumers and credit markets (among other factors). Alongside this interest in market dynamics, EM recognizes the transformed role of the nation-state through what Jänicke and Weidner have called 'political modernization', a process marked by decentralization and flexible and conciliatory governance procedures. This type of governance actually benefits from the incorporation of social movements in institutional decision-making processes (a position that radically diverts from the earlier marginalization of these movements). The government, in this airing, appears to match the role played by Beck's 'subpolitics', a byword for the increasing role by civil society groups in round-table policy consultations, for 'conflict-resolution without state interference' (Mol, ibid.).

Jänicke and Weidner (Jänicke and Weidner, 1997; Weidner, 2002; Weidner and Jänicke, 2002) have produced a strong model that combines the ingredients composing the above clusters. This facilitates the cross-national investigation of capacity for EM that has very much informed the investigation performed in Chapters 7 and 8 on the EM capacity of Olympic hosts. Jänicke and Weidner's model has already been applied in their comparative study about the capacity for implementation of environmental policy within 30 developing and developed countries. According to this model, capacity for EM is constituted by the following components:

1. Actors: the supporters and opponents of environmental issues, as well as their strength and competence to act depending upon existing structural conditions.
2. Strategies: Actors employ a range of strategies in order to utilize their existing capacities; these strategies depend upon existing knowledge and the possibility of tactical and coordinated action.
3. Structural Framework: the opportunity structure that is available to the protagonists of environmentally-related actions. To facilitate better analysis, this variable can be broken down into conditions that emanate from different political, economic and cultural contexts.
4. Situational Context: Actions can also be conditioned by short-term, alternating situations, such as major environmental incidents like Bhopal or Chernobyl. The media's construction of these incidents as seriously harmful to the health and lives of the people who were exposed to them was paramount in increasing the leverage of environmental groups and other advocates for environmental protection. It is worth noting that in periods of economic recession, economic and administrative functions operate under more restrictive conditions.
5. Character of the Problem: What is the urgency of the problem? To what extent is it easy to apply a solution? The answer to these questions is also assessed based on the visibility of harm and the extent to which the gravity of a given will have an impact on future generations. The economic clout of the polluter plays an equally important role, especially when it entails important societal leverage and affected groups are disaffected and weak.

By using this five-component model, Jänicke and Weidner not only added to the five, predominant EM clusters, they provided a compact and inclusive means for assessing the environmental policies of different nation-states that was true irrespective of those states' developmental indicators. The variables within the present study demand a systematic, interdisciplinary understanding of the societies under consideration. Aside from using EM as a guiding theory and method of analytical assessment, Jänicke and Weidner's model will be useful in describing how changes that facilitate environmental reform occur. The following section, therefore, thinks through the implementation of EM in both national and international policy-making while at the same time casting an eye toward the preconditions that are necessary for the facilitation of this process.

EM in action

The creation of the EU can be seen as a key factor in the diffusion of environmental policy across Europe. Johnson (2004) claims that the incorporation of the environmental dimension into all European political sectors was one of the main ambitions of the Fifth EC Environmental Action Programme (1992–2000), which came into force 1 January 1993 and established the EU legislative agenda for the decade that followed. (It was later incorporated into the Treaty of Amsterdam in 1997.) An assessment report on the achievements and failures of the Programme notes that

> [e]nvironmental protection has moved forward in the Community, and Community policies have brought about a reduction in the trans-boundary air pollution, better water quality and the phasing out of substances which deplete the ozone layer. This progress has been somewhat modest because the Member States and the various sectors covered by the programme have not really managed to take proper account of environmental concerns or to integrate them into their policies. The Union is still far from having achieved the broader objective of sustainable development laid down in the Treaty of Amsterdam.
>
> (Europa, 1999)

In July 2002, the European Parliament and the Council laid down the Sixth EC Environmental Action Programme (Europa, 1999), which took into consideration the significant enlargement of the EU. At this stage, in order to complement the discussion of the policy implementation of EM, I'm interjecting with a discussion on the interrelationship between two core EM notions', economic growth and environmental preservation.

As expected, there has been scepticism about the feasibility of a meaningful compromise between economic and environmental targets – especially as such a compromise might be applied to developing countries. In response to this scepticism, supporters of EM sometimes offer their interlocutors a theoretical proposition made by economist and Nobel laureate Simon Kuznets. According to Kuznets, as low-income countries begin to develop and become middle-income countries, living standards and income inequality in these countries temporarily worsen, but 'continued per capita income growth after middle-income status has been achieved tend[s] to be followed by lower levels of income inequality' (Buttel, 2003, p. 326). By extending the working hypothesis that guided Kuznets's research and by replacing the concept of 'income inequality' with that of a healthy environment, we can draw similar conclusions about development and sustainability. Thus,

> as countries move from low income to middle income, their impact on the environment will become increasingly negative or destructive because of the expansion of production, their still-growing population, and the inefficiencies associated with obsolete production practices and equipment. But as the development process proceeds to a high level of per capita income, one can expect that environmental performance will progressively improve due to private incentives to dematerialize, to the fact that environmental movements will increasingly organize to address environmental concerns and risk, and because governments will improve their capacity to militate against environmental degradation.
>
> (Ibid.)

This idea of uneven development and environmental degradation will be important to bear in mind when we engage with the Chinese and UK cases. For the time being, it is important to point out that the dynamic of EM, as a theoretical foundation for feasible solutions to environmental problems, is neither diminished nor increased by proving or disproving the validity of Kuznets' hypotheses.[1]

Johnson (ibid.) adds that many studies on the Kuznets curve indicate that its relevance is better manifested in local issues of environmental degradation (e.g. the pollution of urban atmosphere, potable water contamination) rather than more global issues like GHG emissions. In particular, she (ibid., pp. 158–9) advances two related propositions:

Kuznets curves may not apply to developing nations because of the industrial flight of polluting industries and the inability by these nations to relocate their own polluting industries.

On the other hand, developing nations may be able to fulfil the requirements of environmental Kuznets curves by using technologies that were not available in developed countries when they were at the same stage in their own developmental processes. Johnson expands on these issues by arguing that economic development is an essential but insufficient condition for the reduction of pollution levels. The objective of a reduction in levels of pollution has been much easier to achieve in countries with democratic institutions; indeed, the World Trade Organization has argued that, among countries with similar income levels, environmental deterioration tends to be worse where income inequality is higher, basic education is lower, and political and public freedoms are limited (ibid.).

EM places great emphasis on investment in and use of technological innovation in order to facilitate environmental protection, while at the same time paying attention to competitive principles and market requirements. The underlying idea behind this perspective is that stricter environmental regulations and policies do not lead to additional costs for businesses but instead create the motivation to innovate and compete. For Johnson, this is evident in the so called 'Porter Hypothesis', which claims not only that there is compatibility between economic growth and environmental protection but that competition is itself dependent upon this connection (see Porter and Linde, 1995).

Far from standing in opposition to efficient and competitive markets, EM actually suggests that market dynamics can play a crucial role in environmental reform (albeit under regimes that allow for intervention in cases of market failures). In fact, EM demonstrates that policy-making institutions appear to be veering toward the role of market enablers and guardians through the use of new economic instruments such as green taxes, eco-labelling, and Emissions Trading Schemes (ETSs). As Johnson (ibid., pp. 164–5) writes,

[i]n an increasingly integrated world, the configuration of trade, investment and production chains is determined by market criteria

rather than being constrained by borders which can often be sub-optimal from an economic perspective. Ecological modernization reflects this trend. However, governance structures often lag behind market reality. In environmental terms, this is certainly true both globally and at the European level. However, partly because integration is more advanced and more planned at the European level, EU environmental policy arrangements and institutions are more developed than supranational policy and institutions or the Multilateral Environmental Agreements that have proliferated in recent years [...]

Despite the obvious differences, there are fundamental similarities at play in the Europeanization and globalization process. EM ideas are strongly represented in EU policy. They are present also, albeit less consistently and less strongly, at a global level. However, because of the similarity of forces at a European and international sense, and as multinational enterprises adapt EM ideas and practices, and assuming the environmental Kuznets curve effect kicks in, there is scope for greater spread of EM practices on a more global scale.

Indeed, although EM is strongly represented in EU policy, it would be misleading to suggest that there is cross-national uniformity in its application. Environmental concern differs between EU member states. This factor is of particular importance when we take into consideration the fact that two of the countries studied here, Greece and the UK, are EU member states that have also been facing the impacts of a severe global economic crisis.

How can national EM capacity be measured in this vexing context? As it has been argued by Hannigan (2006, p. 26), the two nation-states (Germany and Holland) where EM had originated still dominate the discussion on EM and 'little is said about the social and political barriers that are likely to be faced in trying to implement [EM] strategies'. This has been taken into consideration by Frijns and colleagues (2000, p. 258) in their study of EM potential in non-Western European countries where they made use of the eight essential social institutions for the study of ecological restructuring in Western Europe.

(Mol, 1995)

Thus, essential institutional characteristics for the development of EM capacity include:

1. a democratic and open political system;
2. a legitimate and interventionist state with an advanced and differentiated socio-environmental infrastructure;
3. widespread environmental consciousness and well-organized ENGOs that have the resources to push for radical ecological reforms;
4. intermediate or business organizations that are able to represent producers in negotiations on a sectoral or regional basis;
5. experience with and tradition in negotiated policy-making and regulatory negotiations;
6. a detailed system of environmental monitoring that generates sufficient, reliable and public environmental data;
7. a state-regulated market economy that dominates production and consumption processes, covers all the edges of society and is strongly integrated in the global market; and
8. advanced technological development within a highly industrialized society.

Cohen (1998, p. 149) further identifies 'cultural capabilities pertaining to societal commitments to science and environmental consciousness' as key parameters that can impact upon and influence these institutional characteristics. In essence, these policies are consistent with EM. For this reason, the cultural dimension plays a key role in the examination on the EM capacity of host nations that follows.

Challenging EM claims

York and Rosa's (2003) critical review of EM has challenged EM's core assertion that modernized countries have entered a process of dematerialization. York and Rosa suggest that EM theorists 'be much more precise in their specification of their argument that ecological modernization reduces energy and resource consumption, because although in some cases the resource of energy use per unit of production decreases in modernized nations, total resource and energy consumption typically increases in such nations'. Given this situation, EM theorists 'must either demonstrate that modernisation does in fact lead to reductions in energy and resource consumption [...] or [...] dramatically scale back their claims regarding the supposed beneficial effects of modernization on the environment' (ibid., p. 281). In a similar vein, I (Karamichas, ibid., p. 170) have argued elsewhere that 'a successful bid to host the Olympic Games may affirm that the environmental standards prescribed by the IOC will be met, but it means that neither these standards will be implemented as prescribed nor the successful implementation of environmental standards for the projects associated with the Games will inevitably lead to a general ecologization of the national economy'. This study's exploration of four different Olympic editions is largely an attempt to test these issues.

Concluding remarks

After a detailed presentation of both the theoretical and policy aspects of the EM perspective, including its close affinity to SD, the chapter singles out those indicators that can help in assessing the validity of the claims made by EM in specific national contexts and, by extension, how it can be used to assess the ways in which hosting the Olympic Games may lead to the development of the EM capacities of host nations.

The objective of this chapter was to argue that both dimensions of reflexivity theory, the risk society and EM, can be employed as useful frames for the study of the environmental problematic, as well as contribute to the development of an effective environmental policy framework. The chapters that follow use reflexivity in order to test the claim that hosting an Olympic edition can help the host nation to develop environmental sustainability (ES), which is to say the EM capacity of the host nation.

Because the two different facets of reflexivity are based on developments that have taken place in advanced western European countries (Germany and Holland), it has been necessary to complement these perspectives by developing an investigative framework that also takes into account the cultural parameter and its impact on the understanding of the environmental problematic within the existing institutional frameworks of different countries.

Returning to Hajer's famous dictum (1996, p. 252) – 'The hardware can be kept but the software should be changed' – it is necessary to decipher the nature of 'hardware' in applied different national contexts. If 'hardware' simply means the established political and economic modernization framework, it is essential to examine how well the nation under examination fits within this framework. After all, EM advocates that the solutions to environmental problems lie in the progressive modernization of societies (rather than in the 'de-modernization' or 'counter-modernization' that is advocated by radical environmentalists) (Buttel, 2000a, p. 29).Changes to a nation's 'software' that are the result of its hosting an Olympic edition will be meaningless if the nation's hardware is still permeated by characteristics that are not conducive to the modernization process. The same phenomenon can be found in periods of uncertainty with significant impact upon the established patterns that the national hardware has been programmed to facilitate. Both scenarios will materialize in the following examination of the post-Olympic EM capacity of four host nations. Before embarking upon this detailed examination, however, I want to discuss the appearance of the environmental dimension in the IOC's process for awarding the Games.

6
The Greening of the Games

Having considered the history of environmental concern in previous chapters, this chapter examines how the International Olympic Committee (IOC) has taken environmental issues into account in its operations. The chapter demonstrates that the incorporation of the green dimension in the planning and organization of the Games took some time to develop alongside the rise of environmental concern since the 1970s (and the sociological engagement with the environmental issue). Nevertheless, the green dimension has now been fully incorporated into the planning and staging of successive Olympic editions. Starting with Sydney 2000, that development has also been accompanied with plans for the long-term legacy of the Games. These long-term legacies include the ecological modernization (EM) of the host nation. More recently, these calls have led to development of Olympic Games Impact (OGI) studies.

Olympic Games and environmental concern

Despite the rise in environmental concern among Western publics through the 1960s and 1970s, the IOC has been extremely slow in adapting its procedures to take this into account when awarding the Games. For instance, public referenda held in Denver turned down the IOC's offer to host the 1976 Winter Games on the basis of environmentally destructive practices (Lenskyj, 2000, p. 157). It took two decades for the IOC to make the environmental dimension a necessary requirement for awarding the Games.

Since the first modern Olympics in 1896, Olympic Game hosting has been, in most occasions, an extremely costly affair with few tangible benefits for the host nation. Indeed, 'after the Second World War, "the cost of hosting the Games became a prohibitive factor in cities bidding for the Games", especially "before the advent of substantial revenues from major corporate sponsoring and broadcasting rights"' (Short, as cited in Close et al., 2007, p. 10). In order to illustrate this point, Short (2003) claims that '[t]he 1976 Montreal Games were responsible for exposing the gulfs that had opened up between a host city's "programmatic aspiration", the "high cost" of staging the Games, and "actual revenue". After taking into account "the cost of all infrastructural investment", the city suffered an [immense financial] loss falling to local and regional taxpayers. Montreal's experience fed the reluctance of other cities to bid for the Games, until Los Angeles "charted a new course"' (ibid.). It is reasonable to assume that the environment did not feature prominently in the list of priorities of an Olympic host city.

Yet according to Toyne (2009, p. 232) 'nearly all Games by their sheer scale have considered how to manage their impact. For example, host cities such as Rome (1960) and Montreal (1976) implemented changes to their urban transport systems that reduced car use and provided an improvement in air quality. The Tokyo Games in 1964 provided Japan with an opportunity to tackle its capital's environmental problems'. However, these benefits are connected to infrastructural changes, which – irrespective of their many needed improvements to the urban environment of the host city – are of uncertain longevity and aptitude in facilitating environmental sustainability. After all, as Evans (2007, p. 298) has noted 'analyses of long-term regeneration effects are notable by their absence. Olympic effects are subsumed into wider redevelopment and competitive city narratives. This makes it problematic to measure the true impact of the Games, which become a symbolic but simply temporal event in a city's evolution and another chapter in the Olympic history book'.

It is important to highlight the fact that Los Angeles was the sole bidder for the 1984 Games. As such, it 'had a strong negotiating position with the IOC' and crucially the US' National Olympic Committee (NOC) 'set up a private non-profit corporation', the Los Angeles Organizing Olympic Committee (LAOOC), in order to avoid overburdening the city taxpayers (ibid, pp. 10–11). The end result was that LAOOC profited from its involvement. Moreover, sponsoring 'corporation [34 in total, including Coca-Cola, Mars, and Anheuser Busch] achieved global penetration as the Games were broadcast to 156 countries; local businesses made money and the city became the center of world attention without accruing long-term costs or heavy debt burdens' (ibid.; see also Andranovich et al., 2001; Burbank et al., 2001). This was clearly a pivotal moment, as it confirmed the Olympic Games as a money-making enterprise. Yet this development did not inspire confidence among environmentalists. It took some time for them to reduce their scepticism and opposition to Olympic Games hosting; and even more to support the Games by acting as advisors and facilitators. It could be argued that it boiled down to demonstrating that the profitable Games that had emerged in Los Angeles could coexist with environmentalism in a mutually reinforcing relationship.

> The Games (and perhaps other sport mega-events) accordingly operate as showcases for the internalization of environmental values and norms, with a wide mimetic potential in both geographic and public policy sector terms. In other words, high environmental standards are no longer seen as antagonistic to the development of economic growth regimes, becoming instead [following Barry 2005, pp. 303–4] a key source of market innovation and future economic growth.
>
> (Hayes and Karamichas, 2012a, p. 11)

That process was developed sequentially with the institutionalization of the environmental movement in some Western democracies, and the simultaneous emergence and popular acceptance of sustainable development (SD) by governments and business interests. That developmental sequence has paralleled the emergence and development of environmental concern and sociology.

The greening of the games

The actual concern of the IOC with environmentalism can be traced back to 1986, when its president, Juan Antonio Samaranch, declared that the environment was the third pillar of Olympism (sports and culture are the first and second pillar). However, it was the Rio de Janeiro

Earth Summit in 1992 and the growing support for SD that made the ambition of the IOC viable. The Local Agenda 21 (LA21), drafted by the United Nations Environmental Programme (UNEP) for the Summit, was adopted by 182 governments and offered a manual for developing an LA21 that was specific to individual country or community requirements. In 1994 the IOC, in collaboration with UNEP, began to make its ambition for a third pillar more of a reality, and by 1995 the IOC had its own Sport and Environment Commission. This has been exemplified since then by item 13 (of 18) in the IOC's Olympic Charter:

> To encourage and support a responsible concern for environmental issues, to promote sustainable development in sport and to require that Olympics are held accordingly.
>
> (IOC, 2011, p. 15)

In 1996, a paragraph on environmental protection was added to the Olympic Charter, defining the IOC's role with respect to the environment such that

> the IOC sees that the Olympic Games are held in conditions which demonstrate a responsible concern for environmental issues and encourages the Olympic Movement to demonstrate a responsible concern for environmental issues, takes measures to reflect such concern in its activities and educates all those connected with the Olympic Movement as to the importance of sustainable development.
>
> (IOC, 2007, p. x)

By 1999, the IOC had operationalized its own version of LA21, which called for:

1. improving socio-economic conditions;
2. conservation and management of resources for sustainable environment; and
3. strengthening the role of major groups.

These calls were accompanied by a set of more concrete proposals about how these goals may be achieved during the Games. These proposals included extensive use of solar panels at venues and related facilities, the conduct of EIAs (Environmental Impact Assessments) for related projects and environmentally responsible transportation to and from Olympic venues (G-ForSE, n.d).

Nevertheless, the first practical implementation of environmental concerns took place in the Lillehammer Winter Games of 1994, five years before the IOC put forward its own environmental guidelines. In Lillehammer, grassroots activists mobilized against Norway hosting the Games, animated by the environmentally damaging 1992 Winter Olympics in Albertville and the Savoy region of France. These protests were directed against the Olympics in general and specific projects associated with the Games in particular. Norway was also involved in drafting the UN Commission for the Environment report, 'Our Common Future', which formed the basis of the SD principle. These factors led to consideration of the Games' environmental impact from an early stage and eventually led to the creation of a paradigmatic case of organizing a mega-event with a minimal environmental impact (Cantelon and Letters, 2000; Caratti and Ferraguto, 2012; Lesjø, 2000, p. 290). The following four points were implemented in the planning and organization of the Games and as such kept the environment at the forefront:

1. Companies were instructed to use natural materials whenever possible.
2. An emphasis was placed on energy conservation in heating and cooling systems.
3. A recycling programme was developed for the entire Winter Games region.
4. A stipulation was made that arenas must harmonize with the surrounding landscape.

Although the Winter Games differ substantially from their summer counterparts in terms of their demands on the natural environment, Lillehammer provided a benchmark for the Sydney Olympics in 2000 (Lenskyj, 2000, p. 159). Indeed, Sydney organized the first 'Green Olympics', with positive reviews of its performance by environmental organizations. Since then,

Rogge has sought to consolidate this orientation whilst attempting to find ways of managing the seemingly ever-increasing scale of the Games. On succeeding Juan Antonio Samaranch as IOC president in 2001, he set up a study commission to establish a blueprint for 'good governance', which reported to the 2003 Prague IOC meeting. Chaired by Dick Pound, the commission produced 117 recommendations to manage the size, cost and complexity of future Games.

(Hayes and Karamichas, 2012a, p. 9)

The recommendations made by the commission indicate strong support for compact and sustainable Games. The recommendations also include 'the preference for temporal installations over permanent ones where post-Games use does not justify the latter, and the encouragement of the use of public transportation and car sharing were possible' (Pound, as cited in Hayes and Karamichas, ibid.).

The first signs that these recommendations were put into action can be seen in the Beijing Games. The Sydney Games may have been the first 'Green Olympics', but like the Olympic edition that immediately followed, Athens 2004, Sydney's Games did not escape exorbitant costs. Nonetheless, the success of Sydney underscores the changing attitude toward environmental concerns, as 'the Games (and perhaps other sport mega-events) accordingly operate as showcases for the internalization of environmental values and norms, with a wide mimetic potential in both geographic and public policy sector terms.' In other words, high environmental standards are no longer seen as antagonistic to the development of economic growth regimes, becoming instead [following Barry 2005, pp. 303–4] a key source of market innovation and future economic growth (Hayes and Karamichas, 2012a, p. 11).

In 2007, the IOC and its president Jacques Rogge were honoured as Champions of the Earth by UNEP. In receiving the award Rogge made the following statement:

Since the early 90s the IOC and the Olympic Movement have progressively taken the environment and sustainability into account throughout the lifecycle of an Olympic Games project. The 'Green Games' concept is increasingly a reality. Today from the beginning of a city's desire to stage an Olympic Games, through to the long-term impact of those Games, environmental protection, and more importantly, sustainability are prime elements of Games planning and operations. I am very proud of this and would like to thank the UNEP for recognising these efforts.

(Beijing 2008)

Furthermore, the IOC agreed to an OGI study in 2001. This study 'is designed to evaluate the Games' legacy for the host nation and city against a raft of social, economic, cultural and environmental indicators, hence providing an "evidence base" for measuring the positive societal consequences of the Games for its hosts' (MacRury and Poynter, 2009, p. 304). The study was first introduced into the formal planning requirements for the 2010 Vancouver Winter Olympic and Paralympic Games. London is the first Summer Games host city mandated to carry out the study (ESRC, 2010, p. 6). In effect, successive Olympic editions have considered the sustainability legacy of the Games (see Table 6.1);

Table 6.1 The Games and the environment

1986	IOC president suggests the environment as the third pillar of Olympism
1988	Seoul
1988	Calgary (Winter)
1992	Barcelona uses green designs in the Olympic Village
1992	Albertville (Winter)
1993	Sydney wins 2000 bid on the basis of strong environmental commitment
1994	Lillehammer (Winter): first 'Green' Winter Games
1994	'Environment' becomes the third pillar of Olympism
1994	UNEP and IOC sign agreement of cooperation
1995	IOC forms a Sport and Environment Commission
1996	Atlanta
1998	Nagano (Winter)
1998	IOC decides the Paralympics and Olympic Games will be held in the same host city and venues
1999	IOC adopts the Olympic Movement's Agenda 21, for sustainable games
2000	Sydney hosts first 'Green Games'
2002	Salt Lake City (Winter) plants 3 million trees and introduces composting and recycling
2004	Athens implements a narrow range of environmental commitments
2006	Turin creates a green legacy
2008	Beijing raises the bar for green ambitions

Source: UNEP, 2009, p. 20.

the application of OGGI by Beijing 2008 has marked a new era for the environmental legacy of Olympic Games.

The following chapter revisits Sydney and examines the legacy left by the 2000 Olympic Games. Using indicators from my examination of the EM perspective, I examine the extent to which the organization of successful 'Green Olympics' is a one-off event or a lasting platform that diffuses its standards across the range of economic and technical activities. Chapter 7 examines the same issues in relation to Athens 2004 and Beijing 2008.

7
Olympic Games and Ecological Modernization: Sydney, Athens and Beijing

Introduction

This chapter examines developments in the ecological moderniza-tion (EM) capacity of three Olympics host nations: Australia, Greece, and China. Our primary question, 'to what extent does Olympic Games hosting lead to the EM development of the host nation?', is here brought into focus. In particular, six EM indicators are examined in order to identify the extent to which meeting the environmental standards of the International Olympic Committee's (IOC) coincides with changes to the institutional and policy frameworks of the host nation. These findings partly confirm those made by earlier work on the Australian and Greek cases. The addition of the Chinese case enriches the debate with important findings about the legacy of the Olympic Games, the EM capacity of host nations, and EM theory.[1]

The examination of post-Olympics EM capacity

This study's analysis framework is closely tied to an examination of each phase in the development of a sports mega-event (Hiller, 2000, p. 192). For the purposes of this study, this developmental sequence is composed of the following: the 'pre-event' phase of IOC bidding applications and the preparations to fulfil environmental commit-ments; the 'event' phase; and the extent to which these preparations and changes signified a post-event commitment to environmental sustainability (ES). The rationale and approach used in Karamichas (2012a, p. 152) has been employed. In that study, by subscribing to the view that the Olympic Games, along with other sport mega-events, are quintessentially modern events, I supported the view that they 'encapsulate the essence of the normative claims made by the ecological modernization (EM) perspective: namely that market-oriented regimes of capital accumulation have the capacity to engage with the environmental dynamic and provide sustainable solutions (ibid.). That way, in order to assess the post-Olympic Games hosting capacity for the ecological modernization of the host nation, I followed the rationale employed by Andersen (2002) in his 'discussion of the effects of Europeanization on Central and Eastern Europe' (ibid.). In particular, by closely following Andersen (2002, p. 1396), I adapted his two contrasting hypotheses as follows:

1. In the wake of their respective Games (which were after all awarded to them, at least in part, on the basis of a range of green claims), 'one should be able to identify marked signs of environmental improvement' in the host nations (Karamichas, 2012, p. 152.).
2. To achieve environmental transformation, the effect of hosting the Olympic Games 'depends more on the supportiveness of domestic political processes' (ibid).

The cautiousness that characterizes the second hypothesis is guided by Andersen's view that 'the degree of ecological modernization in a particular country depends on its capacity for environmental reform as fostered and supported by the character of the political and socio-economic reform process' (Andersen, 2002, p. 1396).

The environment and the bidding process

According to its Manuals for Candidate Cities (MCC), which the International Olympic Committee (IOC) has published since 1992, all candidate cities are required to

1. provide descriptions by means of a map and a chart of the local environmental situation and the environment and natural resource systems used by relevant authorities with emphasis on their interaction with the OCOG (Organizing Committee for the Olympic Games);
2. provide 'an official guarantee from the competent authorities, stating that all work necessary for the organisation of the Games

will comply with local, regional, and national regulations and acts and international agreements and protocols regarding planning and construction and the protection of the environment' (IOC, 1996, p. 45, but repeated with small differences in all other MCCs);

3. carry out Environmental Impact Assessments (EIAs) for all venues;
4. describe the OCOG's planned environmental management system (including possible collaboration with environmental non-governmental organizations (ENGOs) and/or their reaction to the Games);
5. describe the application of environmentally-friendly technology relating to the Games;
6. state the plans for minimising the environmental impact of infrastructural projects relating to the Games;
7. state how plans for waste management (including sewage treatment) are expected to 'influence the city and region in the future' (IOC, 1996, p. 46, but repeated with slightly different wording in all other MCCs);
8. explain how 'the OCOG [will] integrate its environmental approach into contracts with suppliers and sponsors, for example, with respect to procurement of recyclable or compostable goods, in recyclable or compostable packaging' (IOC, 2004, p. 88; the most explicit statement on this issue by the IOC when compared to earlier MCCs); and
9. outline plans for raising environmental awareness (Karamichas, 2012a, pp. 154–5).

Since 2001 the IOC has also agreed to an Olympic Games Impact (OGI) study; London is the first Summer Games host mandated to carry out the study. Chapter 8 presents the findings from the OGI study and adds to the critical points that have been raised in relation to both the post-Olympics EM and sustainable development (SD) standing of the UK.

The dynamic application of all these requirements demands cooperation by state institutions, business entities, and civic organizations. These requirements can significantly affect the organizational capacity of the host nation. In relation to the key issue under examination here, it is clear that a country aspiring to host an Olympic edition must also aspire to develop long-term EM.

Environmental protection in successful Olympics bids

In my previous work on the issue, I highlighted the fact that all Olympic Games hosting bids report on the environmental capacity of the prospective host nation. This is accompanied by plans for further developments on that front resulting from the factoring of the environmental dynamic in the planning of the Games. However, in my study of the bids of two successful bidders, Sydney and Athens, I identified a number of striking differences in the language used. I found that in the case of Sydney, 'we see the expression of an objective that seemed to have been planned irrespective of Olympic hosting and, in the [case of Athens], we are confronted more by a statement of ambition and intention without evidence of any substantial planning in that direction' (Karamichas, 2012a, p. 159). Thanks to the reviews of the Athens Games by leading ENGOs, we now know that Athens failed to realize the ambitious proclamation made by the Athens Organizing Committee (ATHOC) that 'the environment will not only be protected, it will be improved' (ATHOC, 1996, p. 52). However, for those ENGOs 'the environmental plans for Athens 2004 closely followed those of Sydney. As such, although ATHOC did not take advantage of the critical points raised by ENGOs over Sydney 2000 to complement its own environmental action plans, they were still good plans. [...] the implementation of these plans was inhibited by the political process' (Karamichas, 2012a, p. 159).

The truth is that ambitious claims of the kind made by ATHOC have been present in subsequent bids. For instance, in the bid made for Beijing 2008, it was claimed that it would 'leave the greatest Olympic Games environmental legacy ever' (UNEP, 2007, p. 26). The available range of data and discussions support the argument that China has indeed fully embraced EM.

Ecological Modernization (EM) in the post-event phase

Through engagement with the core EM literature and the green legacy aspirations of the IOC, the following six indicators for identifying and testing changes in environmental-event phases were selected:

1. average annual level of CO_2 emissions;
2. level of environmental consciousness;

3. ratification of international agreements;
4. designation of sites for protection;
5. implementation of EIA procedures; and
6. ENGO participation in public decision-making processes.

I used these indicators in my earlier study on Sydney and Athens. In that study, and in light of the flawed environmental performance by Athens, I examined the single environmental success of Athens 2004 – environmental awareness and its key role in facilitating environmental improvements in the 'post-event' phase. I highlighted the need to cautiously apply a full-scale comparison based on the identification of one variable testing positive. I claimed to 'remain cautious – despite the typical claims of event organizers – in deducing causality between event hosting and developing EM capacity. A range of key intervening variables, such as economic growth, the environmental policies of the political party in government, and possible cultural inhibitors to modernization, need to be also accounted for' (Karamichas, 2012a, p. 159). In this study, that testing has been re-activated, albeit with adjustments made for the aforementioned variables.

Sydney 2000

We already know that Sydney raised the standards of environmental performance for the Games. As such, in this case the examination of the different phases is important for the fulfilment of this study's general purposes. The examination of the Australian case is also noteworthy because of the country's economically advantageous position as compared to the two other Olympic host nations (Greece, UK) under examination in this book.

According to Cashman (2009, p. 133),

[i]t is important from the outset to place Sydney's Olympic vision and its legacy – what has been promised and what has been realised – in the context of the time. Sydney's Olympic vision was framed in the early 1990s, when legacy was of much lesser importance in Olympic discourse than it is now. Legacy was then a far more informal and haphazard practice. Although legacy was enshrined in Sydney's bid – making it more attractive to the

international Olympic community and saleable to the host community – it was taken as a given which would occur as a matter of course after the Games. Few plans were put in place to implement and evaluate Olympic legacy after the Games and there was not designated post-Games authority to operate in this period.

This last point places Sydney's legacy much closer to Athens 2004. In order to establish a clearer understanding of what was in fact achieved, we should look more closely at Sydney itself.

Sydney, the capital of New South Wales, has a population of four million and is the largest city in Australia. It experienced rapid growth during the twentieth century and became host to waves of immigrants, originally from Britain, then from southern Europe, and lately from Asia. It has a high-rise central business district surrounded by single-storeyed residential suburbs (Toohey and Veal, 2007, p. 237).

The strengths of Sydney's bid for the year 2000 Games were, reportedly, its claim to be the 'athletes Games', reflected in the concentration of most of the sports in two locations (Homebush and Darling Harbour), and the housing of all the athletes in one village adjacent to the main site. Other influential factors were believed to have been the fact that many of the needed sport facilities were already in existence; the apparent strong support from the community; Australia's long and consistent record of participation in the modern Olympic Games (Australia's claim to be one of only two countries, the other being Greece, to have been represented at every one of the Modern Olympic Games); and the environmental component of the bid.

The environment and the bid

The bid makes extensive reference to existing environmental capacity and plans to expand that capacity. Chalkley and Essex (1999b, p. 301) point out that, for Sydney, 'the Olympic plans give an increased priority to ecological sustainability and issues such as energy conservation, bio-diversity and the need to conserve natural resources. Indeed, the Sydney Organising Committee for the Olympic Games (SOCOG) has published a detailed set of "green" guidelines which are intended to govern the design, layout and the construction of the Olympic facilities.' For instance, a statement made in the Sydney bid, regarding the use of new environmentally friendly technologies, highlighted the existing capacity of Australia on this issue as follows:

> New technologies which help protect the environment are being applied to developments in Sydney for the Olympics.
>
> Many of these techniques have been developed in Australia, which is recognised as a world leader in research on the application of new technology to environmental enhancement.
>
> Within New South Wales, waste and water treatment technologies, such as the use of special membranes, are improving the quality of affluent discharge in various applications.
>
> New ocean downfall systems are significantly improving water quality at urban beaches; water supply treatment technologies are considerably upgrading the quality of municipal water for domestic and commercial use.
>
> Australia has led in many areas of solar energy use, and in designing to improve energy efficiency. A prime example of the latter is the flexibility designed into the air conditioning system for Sydney International Aquatic Centre. The system can create a comfort zone for competitors and spectators. The air-conditioning can run in particular zones of the building when required, thus saving energy.
>
> Technologies are applied, as discussed above, to significantly reduce vehicle pollution and introduce more energy efficient fuel sources.
>
> (SOCOG, 1993, p. 70)

Nevertheless, the promoted green standards were not immune from attracting controversy and opposition. Indeed, 'activists denounced what they saw as an opportunistic top-down strategy that failed fundamentally environmental standards but succeeded in creating a perception of success thanks to the "questionable" or bought-in support of media partners' (McManus, as cited in Garcia, 2007, p. 254; see also Hall and Hodges, 1996, and Hall, 2001). Green Games Watch 2000 (1999) identified a number of loopholes caused by particular policies of the New South Wales (NSW) government, which encouraged selective conformity. McGeoch (1999) argues that the organizing committees were rather choosy in the requirements that they followed. Nevertheless,

> [c]onsistently with the strong commitment to environmental impact mitigation set out in the Olympic Bid, comprehensive plans were developed for the Olympic locations: remarkably, for the first time in Olympic Games planning, management and assessment tools for the infrastructural programmes were used. For each of its operations, the [Olympic Coordination Authority] OCA adopted an Environmental Management System based on ISO 14001. Comprehensive environmental requirements for tendering (including the adoption of an environmental policy by contractors) were also instituted. As part of this 'green' tendering process, an LCA (life-cycle assessment) was introduced for the environmental proposals of all contractors, which were then integrated into a general model detailing all the impacts of the Olympic Stadium.
>
> (Caratti and Ferraguto, 2012, p. 114)

Nevertheless, 'the commitment of the NSW Government and the Organizing Committees was driven by time and budgetary constraints, which took precedence over "green" issues' (Luscombe, 1998, as cited in ibid., p. 115) and democratic consultation (Lenskyj, 2002, p. 140).

Adding to these, critics of Sydney's green reputation argue that the post-event phase was not adequately considered. For instance, criticisms point to the fact that the Olympic park was most of the time empty and inadequately serviced by existing transport networks. In general the under-use of Olympic facilities raised concerns about the wasteful use of taxpayers' money (Cashman, 2009, pp. 136–7). Plans for the management and legacy of the post-events phase were belatedly developed, following the creation of an Olympic Park Authority (Cashman, 2009, p. 137; Caratti and Ferraguto, 2012, p. 115). 'But this initiative is still insufficient to address all the environmental implications arising from – above all – the construction of venues for the Games, for which a plan of sustainable post-event use still remains – almost 10 years later – missing' (Sadd and Jones, 2009, as cited in Caratti and Ferraguto, ibid., p. 116).

At the same time, some commentators contrasted the vast sums spent with the failure of the Australian government to commit to the Kyoto process. Indeed, in the words of a leading Friends of the Earth (FoE) figure, 'the huge sums of money involved could be used, perhaps, not on a two-week athletics festival but on dealing with really pressing environmental priorities, such as a national programme to reduce CO_2 emissions or the dumping of sewage to coastal waters' (Chalkley and Essex, 1999b, p. 306). With these arguments in mind, we can now consider a more detailed exploration of how EM in Australia has fared in the years after Sydney 2000.

Ecological modernization in the post-event phase

(i) Average annual CO_2 emissions

Australia's CO_2 emissions are among the highest in the world. In 2007, total emissions were 27.6 per cent higher than 1990 levels, the baseline year for the Kyoto Protocol, and 12.02 per cent higher than 2000 levels (see Figure 7.1); per capita emissions stood at 19.00t CO_2e in 2007, an increase from 18.3t in 2002 (UNSD, 2010). Among the factors accounting for these high emissions levels are the high

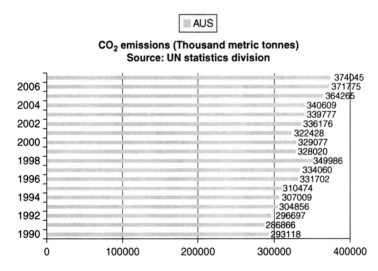

AUS

CO_2 emissions (Thousand metric tonnes)
Source: UN statistics division

Year	Value
2006	374045
	371775
	364265
2004	340609
	339777
2002	336176
	322428
2000	329077
	328020
1998	349986
	334060
1996	331702
	310474
1994	307009
	304856
1992	296697
	286866
1990	293118

Figure 7.1 Australia's CO_2 emissions (kt)
Source: UNSD (2010).

use of coal in electricity generation, the high dependence on car travel for urban transport, and the aluminium-smelting sector (ABS, 2006a, p. 582). As I have pointed out elsewhere (Karamichas, 2012a, p. 160) 'coal is mined in every Australian state and territory, providing 85 per cent of the country's electricity production, and making Australia the world's leading coal exporter – a factor continuously raised during John Howard's Liberal-National coalition governments (1996–2007)'. The government repeatedly refused to reduce the mining and burning of coal for energy use, citing calls for such reductions as attacks on the country's economic interests (Christoff, 2005; Curran, 2009; Doyle, 2010). According to Rootes (2008), this opposition was partially responsible for Howard's defeat by the Australian Labour Party (ALP) in November 2007. Kevin Rudd's administration became mired in internal battles over the introduction of carbon pricing and trading. The ALP government's climate change adviser referred to the problem of climate change as 'a diabolical challenge'. Moreover, he added that 'nations have a collective imperative to adopt their own policies to cut carbon emissions' (Garnaut, as cited in Pietsch and McAllister, 2010, p. 217). These statements gave the impression that Australia remained committed to adopting 'a stronger version of EM'. Nevertheless, 'resistance has continued with industry pressures on the government [...] to weaken its commitment to climate change policy' (Pietsch and McAllister, 2010, p. 220). That perception impacted the political fortunes of Kevin Rudd and ALP, as well as Australia's overall EM capacity.

(ii) Environmental consciousness levels

In a survey 'data linking climate change to the result of the 2007 general election, more than 70 per cent considered climate change to be a "very serious" problem' (Karamichas, 2012a, p. 162). In addition, 'over 60 per cent expressed dissatisfaction with the Howard government's response to the issue' (Rootes, 2008, pp. 473–4). However, Rootes has argued that 'whilst climate change was indeed a very important issue, there is no evidence that it was the decisive issue explaining the election result' (ibid., pp. 479–80). In subsequent developments, such as Rudd's loss of public support over inaction on carbon trading, polling data demonstrated a growing acceptance of the climate change message. In particular, 'a significant proportion (over 60 per cent) [supported] the introduction of an emissions trading scheme (ETS) by 2010, regardless of how the remainder of the international community, including developing countries, decide to proceed' (Curran, 2009, p. 211). In a poll conducted by The Australian National University (ANU) in September 2008, 58 per cent of Australians were clearly in favour of an ETS. Evidence gathered by Pietsch and McAllister (2010, p. 225) suggested that the Australian public was 'reasonably well informed about the ETS and able to form opinions accordingly' (see also Karamichas, 2012a, p. 163). However, according to Tranter (2011, p. 81), these 'analyses were based mainly upon the results of a short opinion poll and did not report on political partisanship or evaluations of political leaders, factors that [he considers] to be crucial determinants of support (or otherwise) for action to address climate change'. Although, in that way we may agree with Tranter and accept that we have possibly missed some important information on certain specificities of Australian environmental concern, the general assessment that we have put here aptly serves the objectives of this work.

The reasons behind 'the environmental concern exhibited by the Australian public before and soon after the 2007 elections, as well as their acceptance of the ETS, range from the unprecedented high temperatures and droughts that preceded the elections to the successful engagement of the public by Rudd on climate change' (Karamichas, 2012a, p. 163; see also Rootes, 2008; Curran, 2009; Pietsch and McAllister, 2010). Yet widespread acceptance of the importance of the climate change issue and the need for ETS by the Australian public in general and ALP supporters in particular (see Tranter, 2011 for a detailed exploration), led to Gillard's replacement of Rudd as the party leader, and the subsequent undoing of ALP's 2007 appeal. This was 'not because a majority of the public had become markedly more sceptical about the threat posed by climate change, nor even because the government's proposed measures failed to command majority public support' (Rootes, 2011, p. 410). Indeed, these developments can be attributed to the fact that these proposed measures, whether as ETS or the Carbon Pollution Reduction Scheme (CPRS) were continuously deferred or postponed. In addition, the 'environment' was considered a 'quite important' issue when 'deciding about how to vote' by 88.3 per cent of the voters, which represented a 7 per cent reduction from 2007 levels (see Figure 7.2). At the same time, 'global warming' was considered a 'quite important' issue when 'deciding about how to vote' by 72.6 per cent of the electorate, which represented

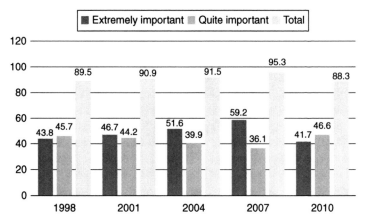

Figure 7.2   Importance of the environment in the 2010 Australian general election
Source: ASSDA (2010).

a 15.7 per cent reduction from 2007 levels (ASSDA, 2009). When voters were asked to rank the 'most important issue during election campaign' in 2010 (out of a choice of 12 issues), 'global warming' was ranked fourth by 7.6 per cent of respondents, after 'Health and Medicine' (27.3 per cent), 'Management of the Economy' (21.6 per cent) and 'Education' (10.7 per cent). This represented a marked difference from the ranking of 'global warming' in the 2007 Australian election study, in which 'global warming' was ranked as the third 'most important problem facing Australia' by 12.0 per cent of respondents.

Gillard entered the 2010 election campaign with a much-derided proposal for a 'citizens' council' to assemble political consensus on climate change. The election cost ALP its lower house majority. This forced the party to seek support from the Greens and three of the four independent cross-bench MPs, who expected substantive policies on climate change to form a minority government (Economou, 2010; Rootes, 2011). In July 2011, the Australian government unveiled a carbon tax plan that was hailed as the 'cheapest means of delivering a "clean energy future" and meeting Australia's target of cutting emissions five per cent by 2020'. Opponents of this proposal argued that it would have a negative economic impact and create job losses. Gillard confirmed 'that the new tax will come into effect from July

[2012], hitting Australia's 500 emitters with a $23 a tonne carbon tax. She also confirmed that the tax will rise 2.5 per cent a year until 2015 when the levy will transition into an emissions trading scheme that is expected to be the world's third national greenhouse gas-and-trade scheme after the EU and New Zealand' (BusinessGreen Staff, 2011). This plan bitterly divided the nation, but was successfully passed as a carbon tax laws on 8 November 2011.

(iii) International treaty ratification

Rudd's engagement with climate change was not the first time ALP embraced environmental issues (see Economou, 2010). Papadakis argues that there was noticeable promotion of environmental concerns by ALP during the 1970s. This promotion of environmental concerns was exemplified by Australia signing a number of International Conventions related to the environment, which led to the establishment of many departments with environmental responsibilities and further facilitated the capacity of individual states for environmental management and regulation (Papadakis, 2002, pp. 24–5). However, Karamichas, (2012a, p. 164) argues that during Howard's successive administrations

> Australia was (alongside the USA) a key country not to ratify the Kyoto protocol, despite negotiating an agreement permitting it to raise its GHG [greenhouse gases] emissions by 8 per cent on 1990 levels by 2010. The rationale was that meeting Kyoto targets was a potential impediment to Australia's capacity to compete in the global market, pointing to the country's competitive advantage in energy as a net coal exporter, job losses, and a reduction of GDP on the one hand, and the small contribution by Australia to global emissions, uncertainty surrounding the human contribution to climate change, and the lack of binding conditions on developing nations.

Australia signed the protocol in December 2007 but during Rudd's tenure the ALP government failed to meet the Australian public's expectations on climate change policy.

(iv) Protected natural site designation

Australia 'has one of the largest and greatest national park systems in the world, covering over 24 million hectares, with such diversity as lush rain forest to waterless desert' (ANP, 2010). Yet the country suffers large forest fires on a yearly basis. In February 2009, 160 people lost their lives in southern Australia in such a fire. For the Australian Greens, the fires were a 'sobering reminder for this nation and the whole world to act and put as a priority the need to tackle climate change' (Walsh, 2009). 'If we accept the argument put forward by the Australian Greens that climate change was the main cause behind these fires, could it be argued that if the Howard government had ratified the Kyoto protocol maybe the preparation for such a possibility would have been more effective?' (Karamichas, 2012a, p. 166). There are a myriad of possible responses to this question. Nonetheless, forest fires are a yearly occurrence, and home-building in fire danger zones has gone on for many years. The severity of the 2009 fires and their death toll were exceptional. In February 2011, forest fires near Perth resulted in the loss of as many as 40 houses, but there was no loss of life, as security forces enforced evacuation measures put in place following 2009's disaster (Astbury, 2011). Despite these devastations, Australia still has an enviable protected natural site designation record.

Olympic hosting did not appear to have any positive impact in relation to this EM indicator. Perhaps this is a harsh statement if one considers the vastness of Australian territory and the interaction between natural and social variables over the occurrence of forest fires. In any case, some may also argue that things could have been worse if the Games had not been hosted in Australia.

(v) Environmental Impact Assessments

The incorporation of environmental concerns with economic objectives was first attempted through the Introduction of the Environmental Impact Statement (EIS) in the 1960s. However, the EIS process fell short of achieving its objectives (Karamichas, 2012a, p. 167), namely, to achieve a balance between economic and ecological conditions. According to Elliot and Thomas (2009, pp. 123–4), it took nearly three decades (until the 1990s) for the adoption of the 'National Approach to Environmental Impact Assessment in Australia' with the principal aim to 'eliminate the prevalent duplication of EIA procedures among different government jurisdictions (state government and the commonwealth government) and improve consistency' (Karamichas, 2012, p. 167). Yet according to Papadakis (1993,

p. 112), successive Australian governments have used EIAs to measure the level of public support for various projects and also to delay
politically uncomfortable decisions.

I have noted (Karamichas, 2012a, ibid.) that according to Elliot
and Thomas (2009, p. 124), '[t]his approach was very much in action
during the planning stages for Sydney 2000'. I also highlighted that
the key aspect of importance is the identification 'environmental
significance [...] as a key factor in determining the need for EIA'
(Karamichas, ibid.). It appears that 'the degree of significance can
have many interpretations, and the use of EIA for the Sydney Olympic
projects is a quintessential example of this, as a range of Olympic-
related projects were authorised without being subjected to the full
EIA process, which was considered to impose unnecessary delays
on a development of national importance' (see Hall, 2001; Hall and
Hodges, 1996). Indeed, the NSW 'state government introduced fast-
track planning procedures in 1995 denying Sydney residents the
right to initiate court appeals against Olympic construction projects,
and its Olympic Coordination Authority (OCA) Act suspended all
projects linked with the Games from the usual requirements of EIS
(Hayes and Karamichas, 2012a, p. 15).

This has been further supported by state-level studies of capacity
to assess the health impact within EIA. The qualitative, descrip-
tive analysis conducted by Harris et al. (2009) in relation to 22
Major Project EIAs in New South Wales (the Olympic state) is a case
in point. Although that study began by acknowledging the leading
position held by Australia, it identified a number of 'deficiencies
in practice and legislative provisions across states and territories'
(ibid., p. 311). These deficiencies were attributed to constraints
emanating from pressures for project approval, and land-use planning processes, among others. Another contributing factor was the
fact that 'in Australia, it is the responsibility of the private developer,
rather than an independent body, to carry out the EIA' (Carter, 2007,
p. 302). 'In effect, little has changed in the post-Olympic phase on this
issue, due to the inherent ambiguity of the EIA process' (Karamichas,
2012a, p. 167).

(vi) ENGO participation

In his review of environmentalism in Australia, Doyle (2005, 2010) has
argued that from the mid-1980s to the early 1990s, some of the most

popular ENGOs were institutionalized through their participation in the decision-making process of the ALP government. With the election of the Howard government in 1996, three years after the Games were awarded to Sydney, this collaborative arrangement was minimized. That period saw the state 'increasingly remove itself from its role as environmental legislator, monitor, and regulator and again [like the 1960s to mid-1980s period] set itself up in active opposition to environmental concerns' (Doyle, 2005, p. 72). In my earlier study on this indicator (Karamichas, 2012a, p. 168), I pointed out that according to Curran (2009), '[p]ost-Howard, there has been an evident reincorporation of ENGOs into policy consultation, but the ALP's apparent dependence on certain economic interests (mining, for example) has already caused significant strain on this relationship'. Since Rudd lost power, and despite Gillard's proposed carbon taxes, it is clear this relationship remains strained. The dependency of ALP on support by the green milieu – as demonstrated by fluctuations in Green Party/ALP support during the demise of Rudd/ascent of Gillard period (see Rootes, 2011, pp. 411–12) – is still an ongoing parameter. Nevertheless, like before, 'if the legacy of ENGO participation in preparation for Sydney 2000 is to go by, there is no reason to dispute that this process might also be characterized by immense selectivity in its inclusiveness/exclusiveness of participant organizations' (Karamichas, ibid., p. 168).

Athens 2004

The Athens 2004 Olympics failed to match the environmental successes of Sydney 2000. Due to the Games' contribution to the environmental awareness of the Greek public, it was deemed appropriate to measure Greece's standing with the indicators used for Australia. In this section, I demonstrate that, along with the political dimension (Karamichas, 2012a), Greece's financial predicament in 2010 has played a pivotal role in the negative aspects of these indicators. As such, Greece's post-Olympic capacity for EM continues to be negatively affected.

According to the 2001 census, the population in the greater metropolitan area of Attica – the Athenian conurbation – was recorded at 3,761,810 inhabitants, which is 34.3 per cent of the total population of Greece. Due to the devastation caused by World War II and the Civil War, in the post-war years, Greece experienced extensive rural immigration to urban areas. This influx in the urban population resulted in unplanned, unregulated physical growth that provided city dwellers with little infrastructure. As a result, Athens exhibits – and this effect was more pronounced at the time of the bid – all the disadvantages of a densely populated urban conurbation in the European periphery: unplanned residential areas on the outskirts, lacking or obsolescent infrastructure, degraded built fabric, traffic congestion, environmental pollution (Beriatos and Gospodini, 2004, p. 192; Panagiotopoulou, 2009, p. 145).

> Following the 1990s and the experience of big international events used by large cities as a catalyst to overcome their spatial disadvantages, to improve urban space qualities and to upgrade themselves in the hierarchies of the global urban system, the 2004 Olympic Games have been considered as an opportunity and a challenge for Athens. Although not explicitly stated by either the state or the Organizing Committee of Olympic Games 2004, different points of view converge in that the strategy underlying Athens' candidacy and the city's preparation for Olympics 2004 was to enlarge the city's development prospects and put Athens on the map as a major metropolitan center in southeast Europe. (Beriatos and Gospodini, ibid.)

Although the historical details underlying each Olympic host may be different, the overall rationale behind the decision to bid for the Games bares similarities across all candidate cities. These rationales frequently boil down to a desire to enhance and strengthen the developmental capacity and growth dynamic of the host nation; this goal often coincides with other objectives, such as environmental sustainability.

That environmental objective is supported by the IOC in upholding and promoting the three pillars of the Olympic Movement (sport, culture and the environment) in awarding the Games. Below, I outline the specifics of the Athenian case in order to tease out the precise aspirations of the bidding nation, and the environment's place in these goals.

These specificities have compelled Panagiotopoulou (ibid., p. 146) to start her treatment of Athens 2004 by stipulating that, '[c] compared with other cities, Athens had two major advantages in being selected to host the Games: its special historic ties to the Olympics and the almost universal support of the Greek people for hosting the event. Over 90 per cent of the total population considered the Games as a constituent part of their cultural heritage and national identity'. I have already commented on the professed continuity of the ancient Games to their modern incarnations in the first two chapters of the book. Most importantly, I claimed that the 'Olympics always occupied a special place in the imaginative construction of Greek national identity. They are among many key components of Hellenic historiography that have been used, since the foundation of the Modern Greek state, to lay claim to a glorious past and in effect to posit the uninterrupted continuity of the Greek nation over the centuries. As such, hosting the Games was a task of immense importance, [...], but it contained intense added value in the Greek case' (Karamichas, 2012c, pp. 163–4). Greece has attempted to host the Games since the nineteenth century (see Young, 2005); Greece did host the first modern Olympics in 1896. Greece was armed with its history as the birthplace of the Olympics and host of the first modern Olympics when it applied in 1990 to host the 1996 centennial Olympics. Those Games were awarded to Atlanta, to the immense disappointment of the Greeks.

However, as I have argued in Karamichas (2012c, p. 165), [a]fter the shock had subsided Athens decided to bid again to host the Games. Indeed, in 1995 the Hellenic Olympic Committee (HOC) canvassed over bidding for the 2004 Games and managed to submit Athens's candidature "five days before the official deadline" (Gold, 2007, p. 269). "Not wholly surprisingly, Athens's final bid document for the 1997 IOC decision bore similarities to the failed 1990 bid, but with less strident tone in the presentation"' (Gold, 2007, ibid.). The bid was successful, but this success has not been solely attributed to an intelligent restructuring of the earlier unsuccessful bid. Observers have also taken into consideration the need for the IOC to fix its image, which at this point had been tarnished by a range of internal scandals (Burbank et al., 2001, pp. 2–4; Pound, 2004). The idea here is that, by awarding the Olympics to Greece, the IOC could wipe away with one swoop the impact of its scandals.

Still, the choice of Athens as an Olympic host still contained significant risks:

> The largest global sporting and mass media event in the world was to be organised by a small country that was not prosperous, and had no pre-existing infrastructure (for example in communication technologies, transportation, sporting facilities, and so on), nor previous experience in organising such major and complicated projects. Moreover, Greece did not have a particularly positive international image in a period of terrorist hysteria, economic recessions and wars in surrounding regions (Kosovo, and later Iraq).
>
> (Panagiotopoulou, ibid., p. 146)

How did the general public react to the success of the Olympic candidacy? According to Dodouras and James (2006, pp. 73–4), at the start of the 'pre-event' period, and notwithstanding expressed concerns 'about the alarming regional imbalances in Greece and massive commercialization of the Games ... Greeks were still in favour of the event mainly because of the potential economic benefits'. However, towards the end of the 'pre-event' period, a year before the event, opinion polls showed that '40% of the respondents were either a little or not interested at all in the event, while only 36% of them believed that it would benefit the host country' (Dodouras and James, ibid., p. 74).

The environment and the bid

The Athens bid did make reference to existing environmental capacity. However, the language used in the bid submitted by ATHOC (1996) differs from language used in the bid submitted by SOCOG. In terms of using appropriate technology, the Athens bid promised the following:

> The projects planned will be incorporated into, and adapted to, general government policy on the protection of the environment and the remodelling of the area.
>
> Completion of the projects will be carried out with the use of environmentally friendly technologies and materials. Contractor companies will commit themselves to the use of such technologies

and materials, for which they will have to submit detailed schedules and lists.

Uses will be planned for all the permanent installations once the Games are over.

Clearly, when the Sydney and Athens bids are contrasted, the key environmental drive of the latter is 'less one of existing capacity than one of cultural change' (Karamichas, ibid., p. 157). For Kazantzopoulos (the environmental manager of the Athens Games),

> ATHOC, in conjunction with the relevant state institutions, was adopting a pioneering approach towards the environmental planning of the Olympic projects in the 'pre-event' phase. Following the adoption of a relevant legal framework (Law 2730/99), a strategic environmental impact assessment of all Olympic related projects was attempted. Furthermore, environmental planning was incorporated into this new legal framework, such that 'notwithstanding a few points which might have been improved, this is an initiative which has not so far been visible in other development programmes'.
>
> (Kazantzopoulos, 2002, p. 111)

In what appears to be suggesting a follow-up to the Lillehammer and Sydney examples, 'Kazantzopoulos continued by setting out the various ways that the environmental dimension had been incorporated into various Olympic construction projects, ranging from the use of environmentally friendly materials to the bioclimatic planning of Olympic facilities. [Most importantly, he highlighted that these were] seen through a bigger objective that [...] was set by ATHOC: the creation of a sustainable environmental heritage' (Kazantzopoulos, ibid., p. 112). I have pointed out that 'one may retrospectively see this as more of an attempt to reassure the IOC, which had already warned Greece back in 2000, after a number of delays had become apparent, that "they might lose the Games if action was not forthcoming" (Gold, 2007, p. 271) than as a realistic forecasting of what lay ahead' (Karamichas, ibid., p. 157).

In any case, good intentions are meaningless if they are not materialized. Characteristically, Greenpeace and WWF produced negative reports on Greece's performance in relation to the environmental promises in its candidacy. In particular, Greenpeace produced a highly critical report, which claimed that Greece, aside from making a few noticeable, albeit long overdue, infrastructural improvements (the extension of the metro network, the construction of tramway and suburban rail lines, etc.), failed to emulate the lessons learned from Sydney 2000 (Greenpeace, 2004a, 2004b). Similarly, WWF also assessed the Athens Olympics based on the Sydney 2000 benchmarks where Athens was rated at 0.77 on a scale of 0–4. That report also selected infrastructural improvements as well as the promotion of environmental awareness as the only items deserving high score (WWF-Greece, 2004).

Moreover, although the Olympic bid requirements were promoted as a significant component and the precedent of Sydney had show-cased its value, consultation and collaboration with ENGOs was minimal or absent from the Athens Games. The comments made by WWF-Greece (2004, p. 10) on this issue are revealing:

> The reality of Athens proves that the 2004 Olympics host city has not learnt anything from the Sydney experience. Collaboration with NGOs has been particularly problematic. The bid file was designed without any participation of NGOs or other interested groups or individuals, whereas the crucial issue of site selection was decided without any consultation with the citizens of Athens. During the course of the preparations and after bitter conflicts, the ATHOC invited some organizations to collaborate in promoting a green profile for the games. This, however, was not at all welcomed by most NGOs since the entire preparation process was covered in secrecy. As a result, considerable reaction was caused against most construction projects, such as the case of the Rowing Centre of Schinias, the Ping-Pong Centre at Galatsi, the Olympic Village and the Press Centre at Maroussi.

Elsewhere, I argued that a 'great number of Olympic related projects were implemented between 2001 and 2003. Because they were long overdue and in close proximity to the opening of the Games, public consultation was completely sidestepped' (Karamichas, 2012c, p. 167). Karamichas (2005) and Kavoulakos (2008) have identified that factor as an explanatory variable behind social contestation over some Olympics-related projects. Before making use of its oppressive apparatus to quell protest, the Greek state invested in calls for unity in successfully meeting an issue of national importance and that discourse appeared to have the desired effect during the Olympic year. As Kavoulakos (ibid., p. 408) discovered in his study of protest activities related to the urban environment, '[i]n 2004 not only were almost all the Olympic projects completed, thus immediately decreasing the number of possible contentious issues, but there is also a general stance of consent in light of the risk of attracting international negative publicity'.

Moreover, in relation to those projects that attracted significant contestation during the 'pre-event' phase, I highlighted the fact that the file submitted to the IOC in its bid for the 2004 Games was very much a replica of its failed bid to host the 1996 Olympics (Karamichas, 2012c, p. 168). According to Totsikas (2004, pp. 64–7), in neither bid there was a harmonization with the Athens Regulatory Framework (ARF), a vision of the city that was supposed to comply with SD.

Among all of the projects mentioned by WWF-Greece and contested by local citizen initiatives and environmental groups, we find that the Olympic Village was contested from the very beginning. Critics argued that an already demographically challenged city was planning to accommodate a community of more than 15,000 residents and expand into Natura-protected, partly forested and partly agricultural land in Mount Parnitha. In relation to the legacy of the Olympic Village, undoubtedly one of the most notable projects that characterize the undertaking of an Olympic Games, I argued (Karamichas, 2012c, p. 171) that '[t]he legacy of Sydney's Olympic village was incorporated by means of knowledge transfer from Sydney to ATHOC'. That way, the initial plans for the village included as 'necessary preconditions, among others, the implementation of bioclimatic design, the use of renewable energy sources, avoiding using PVC material, the use of energy and water saving technologies' (Telloglou, 2004, p. 104). Nevertheless, we now know that the Athens Olympic Village 'was also added to the pantheon of unfulfilled ambitions that characterized the Athens 2004 Olympics' (Karamichas, ibid., p. 171).

In relation to the social legacy of the Olympic Village, two antithetical viewpoints were expressed at an international conference held in Athens in 2009, one highlighting a positive social impact – affordable housing (see Asimakopoulos 2009) – and the other, the complaints made by village residents about the lack of and/or inadequacies of existing structures and facilities (see Lourdis, 2009).

Other highly contested projects related to the selection of Schinias for hosting rowing, canoeing and kayaking as that area was seen to be of particular ecological and historical importance, due to its proximity to the ancient battlefield of Marathon (see Gold, 2007, p. 274).

In particular, the supporters of that project – the Greek government, ATHOC, IOC and Olympic Real Estates S.A. – maintained that hosting these facilities at Schinias would benefit local ecosystems, and increase prosperity in the local community and developmental growth in the Greek capital. Incompatible infrastructures (airport, motocross, piste, etc.) would be removed, whilst landscape and soil, long suffering from toxic waste, would be rejuvenated. At the same time, these developments were to be supported and further complemented by the establishment of Schinias national park with an institutionally appointed managing body. This managing body would have the core aim of supervising and monitoring land use as well as protecting the area from non-sustainable exploitation of natural resources. For the other side, an alliance of ENGOs (WWF-Greece; Hellenic Ornithological Society; The Hellenic Society for the Protection of the Environment and the Cultural Heritage; The Hellenic Society for the Protection of Nature), the proposed infrastructures not only were incompatible with the protection of a significant ecological area but also likely to have a highly negative environmental impact (see Apostolopoulou, 2009, pp. 233–6). This polarity, typical of developmental projects when the environmental factor enters the debate, is evident in both cases. For one side – though not necessarily the most radical representative – the environmental credentials of the Games were flawed from an early stage. This was not the IOC's viewpoint when the Games were awarded to Athens. In contrast, Schinias was presented as one of the main reasons that Athens qualified as a 'deserving heir of Lillehammer and Sydney on the ES front' (ibid. p. 216).

In following the analytical structure that was used on the Australian case, this section should have moved on to an examination of all six of the identified EM indicators. These indicators have already been examined in Karamichas (2012a). However, that examination was completed in 2010, six years after the Athens Games, at a time when the severity of the economic crisis had just become apparent. As such, before embarking on a full-scale re-examination of the post-Olympics environmental capacity of Greece in relation to the six indicators, I must include an account of the social and political impact of the crisis. I have selected two highly informative articles to assist in this process: Pagoulatos (2010), on the importance of stimulating green growth; and Malkoutzis (2011), on one year of austerity measures.

The former article is well situated at the intersection of the green/EM promises made by Papandreou's PASOK (Panhellenic Socialist Movement) and the grim unfolding of the crisis. Although the repercussions of the crisis had been felt elsewhere since 2008–2009, in Greece the crisis was not yet fully apparent. Indeed, when Pagoulatos (2010) wrote about the state of the Greek economy he drew links between the Olympics and the environment. This approach was marked by identification of the cost of hosting the Athens 2004 Olympiad as a major contributor to Greece's financial woes in 2010, and the promotion of a green economy to stimulate growth and find a way out of the crisis. According to Pagoulatos (ibid., p. 2),

> The 2008–2009 global financial crisis has taken a heavy toll on the Greek economy, albeit in a different manner from other EU economies. Although Greece's GDP has been contracting since the fourth quarter of 2008, recession in 2009 was relatively milder compared to the larger negative average growth rates of both the EU-27 and the Eurozone [...] The global crisis has exacerbated the long-standing structural weaknesses of the Greek economy.

When the PASOK government came to power in October 2009, the deficit that was estimated by the previous New Democracy (ND) government was about 6 per cent of GDP. However, because of a shortfall in public revenues and expenditure overruns, 'the budget deficit for 2009 officially closed at 12.7 per cent. In January 2010, the Papandreou government submitted to Brussels an ambitious three-year Stability and Growth Programme, aiming to bring the general deficit down to [...] below three per cent by 2012' (ibid., p. 2).

This can be seen as a nodal point, during which time an array of speculative attacks in the bond market and a continuous wave of harsh austerity measures were unleashed. This continued unabated throughout 2011 and may continue for the majority of 2012. At the time of writing, the situation is in many respects fluid and unpredictable. For this reason, re-examination of the six EM indicators might be of crucial importance but the findings are more tentative than ever; it is necessary to situate the EM ambitions of the Papandreou government in the context of the austerity measures. Following Pagoulatos (2010), it appears that these ambitions were not perceived as obstacles, and instead were seen as essential stimuli for the country's growth and development.

> Green growth and development occupy pride of place in the Papandreou government's Stability and Growth Programme [...] The green economy agenda is the first among the main policies aimed at enhancing economic growth and employment. The Programme reiterates the commitment to green growth and development as a 'major priority for the country', given the need to address the challenges of climate change and the country's unexplored potential in renewable energy development.
>
> (Ibid., p. 6)

The rationale remained much the same as PASOK's programmatic declarations:

> Green development serves the need for a new growth model for a green economy that has relied far too heavily in the past on domestic consumption financed by bank credit, leading to vast volumes of imports and constant growth of the sheltered sector of the economy at the expense of the export sector. Greater focus on an export-oriented growth model is now being championed in a number of quarters, exploiting the country's comparative advantages: the climate and the sea, which make quality tourism the country's 'heavy industry' (as it is often called); while shipping and financial services, the agro-food industry, health and quality of life services and foreign language higher education have often been indicated as growth sectors the economy must tap into.

> Green development is highly compatible with these objectives and – through synergies with the above – suggests a distinct area of comparative advantage in its own right.
>
> (Ibid., p. 3)

However, by March 2010, with the first supplementation of the stability plan with additional austerity measures, Pagoulatos (2010, p. 7) argued that 'Greece's financial fragility and recession in 2010 imply that both the budgetary aspect and the job-saving dimension of any green development policies will tend to predominate. This could end up the speed and effectiveness of transition to a green economy'.

The latter article is a thorough appraisal of the social and political impact of one year of austerity measures. Malkoutzis begins by presenting a range of offensive remarks, typically based on unrepresentative cases, which were made by populist media in some northern European countries.

> This focus on the more extreme aspects of the Greek society is unnecessary and counterproductive at a time when Greeks are facing grave problems, such as an economy in its third year of recession, an unemployment rate at its highest for more than a decade, tax rates and social security contributions that are among the highest in Europe and constant speculation about whether their country will restructure, default or even return to the drachma. This uncertainty has been compounded by the recent political uncertainty created by failed attempts to achieve consensus between the government and opposition parties, culminating in Prime Minister George Papandreou conducting a Cabinet reshuffle on 16 June after reportedly having offered to resign to pave the way for a government of national unity.
>
> (Malkoutzis, 2011, p. 1)

It is that particular reshuffle that signified the swan song of most, if not all, of the positive projections that were made with the coming of George Papandreou's PASOK for EM capacity in Greece. It is important to bear in mind that the parliamentary ratification of the interim programme attracted a lot of civil contestation that also featured heavily in the global media, albeit mainly when they were marked by scuffles between a minority of masked individuals and the police. Aside from the fact that the public tended to feel angry and hurt by the impact of the austerity cuts and the aforementioned offensive remarks, there was a general perception that the Greek public had already made many sacrifices.

> In 2010, Greece reduced its public deficit by five per cent of GDP, which the Organization for Economic Cooperation and Development (OECD) said was the largest such reduction in a single year by any of its members in the past 25 years.
>
> The Greek government has found it particularly difficult to tackle the structural aspects of the country's economic problems. [...] As a result the weight of the fiscal reform programme has fallen almost exclusively on Greek taxpayers. For instance, one-third of the savings in the public sector in 2010 were achieved by reducing wages, pensions and social transfers. Roughly three-quarters of the increase in revenues [...] came as a result of rises in indirect taxes and social security contributions.
>
> (Ibid., p. 1)

This assessment, based on one year in crisis, came out in June 2011. At the time, the unemployment rate was 'the highest since Greece joined the Eurozone' (ibid., p. 2); there had been substantial wage cuts in both the public and private sectors, which resulted in retail trade decrease and great strain placed on the country's social security system; there had been a marked increase in homelessness and the number of people receiving food and clothing at the main canteen in Athens. Moreover, those NGOs covering for state limitations in social care had been facing a serious prospect of 'closure due to dwindling funds' (ibid., p. 2).

The public reaction to these extremely negative developments was generally marked by a new wave of civil contestation that was markedly different from earlier protests. What clearly stood out was the 'Indignants' movement that first appeared in various squares across the country on 25 May 2011. It was this movement that, leaving aside some populist overtones, demonstrated a change of attitude toward clientelism, nepotism and the perennial corruption of some public sector services. Attitudes toward the two political parties, PASOK and ND (which have alternated power since the transition to democracy) have also changed.

Since the ratification of the interim programme in the Greek parliament there has been a continuation of severe austerity measures, failure to deal effectively with tax evasion, and social contestation. The intensity of some protest action at the annual parades held on 28 October (a holiday commemorating Greece's resistance during World War II) led Papandreou to announce plans to call a referendum on Greece's membership of the Eurozone; this announcement caused immense upset to market trading and great annoyance to the European leaders who only a few days earlier (26–27 October 2011) had negotiated an agreement to slash Greek debt in a European Summit. As such, the referendum plans were abandoned, and a tripartite coalition government – composed of members of PASOK, ND and the far right wing LAOS (Popular Orthodox Rally) with technocrat Lucas Papademos in the prime-ministerial position – was formed on 11 November 2011 with the main task of allowing the EU bailout to proceed and pave the way for general elections. Within this context, the next section deals with post-Olympic Games hosting EM capacity as well as the way that EM fares under conditions of immense financial and political strain. From the presentation of EM offered in Chapter 5, it is clear that this situational context is extremely inhospitable.

Ecological modernization in the post-event phase

(i) Average annual CO_2 emissions

In the Greek case, according to UNSD data, in 2007 Greece's per capita CO_2 emissions stood at 10.2t, with overall emissions having increased by 34.9 per cent from 1990 levels (UNSD, 2010). Through consultation, a report by the European Environmental Agency (EEA, 2008, pp. 134–5), I argued (Karamichas, 2012a, pp. 160–1) that 'Greece recorded one of the largest absolute increases in GHG emissions between 1990 and 2006 in EU15 of 24.4 per cent, set against both an overall emissions decrease of 2.7 per cent by EU15 over this period, and an overall EU-27 target of a 20 per cent reduction in GHG emissions by 2020 on 1990 levels set by the United Nations Framework Convention on Climate Change, and an aspiration to reduce emissions by 30 per cent on 1990 levels. Per capita emissions (tCO_2e) increased by 15.8 per cent over the same period (as against an EU15 performance of –8.4 per cent). Clearly, this is long term data,

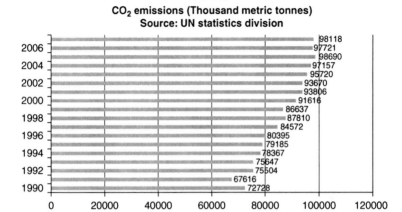

Figure 7.3   Greece's CO$_2$ emissions (kt)
Source: UNSD (2010).

with emissions fuelled by increased transport activity and energy demand, more recently partly counterbalanced by the use of natural gas and hydropower, and is furthermore within Greece's Kyoto allowance of a 25 per cent increase in emissions'.

Interestingly enough, the available data pointed to a decrease by 1.19 per cent in CO$_2$ emissions in 2006 (see UNSD, 2009). However, later on, in April 2008, Greece was accused of inaccurate reporting of carbon emissions and suspended from the UN's carbon trading scheme (Psaropoulos, 2008).

That has now been corrected on the data offered by the UNSD (2010) (see Figure 7.3).

(ii) Environmental consciousness levels

Greeks expressed the highest levels of professed environmental concern in the EU for most of the 1990s (Karamichas, 2003). However, in light of the perpetuation of environmentally harmful practices at both the local and national levels and the public's lack of knowledge about the cause and effect of global environmental issues (see Karamichas, 2003; 2007), the genuineness of that concern was disputed. It is interesting to examine the Greek public's concern exhibited over a global environmental issue like climate change. In fact,

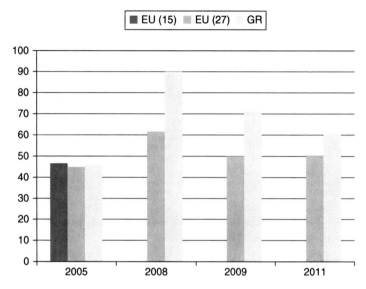

Figure 7.4   Expressed concern about climate change – EU and Greece
Source: European Commission (2011).

the highest issue of concern for the Greeks in 2008 and 2009 was climate change (see Figure 7.4), with 71 per cent of the public seeing it as the 'most serious issue currently facing the world as a whole' after 'poverty, lack of food and drinking water' (72 per cent) and 'a major global economic downturn' (68 per cent) (European Commission, 2009). In an earlier discussion of these findings, I argued that 'there is good reason to believe that the Greek public's concern on climate change is much more sincere and better informed during the first decade of the 2000s than in the 1990s. It is likely that this can be attributed to international factors, such as the promotion of the role that human activity has had on climate change since the 2007 Nobel peace prize was shared by the IPCC and former US vice-president Al Gore, and to national factors such as the extremely devastating forest fires of summer 2007, rather than the staging of the 2004 Olympics and the promotion of environmental awareness associated with them' (Karamichas, 2012a, p. 163).

Nevertheless, this view was influenced by Papandreou's PASOK and its programmatic declaration that highly promoted the principles of EM. Although I did not initially think that concern about climate change expressed by the Greek public was likely to have changed much since 2008/9, the findings of the 2011 special Eurobarometer on climate change (European Commission, 2011) may demonstrate how the impact of a severe economic crisis has downgraded that concern when compared with 'bread and butter' issues. For instance, in the 2011 special Eurobarometer on climate change (European Commission, 2011), the percentage of Greeks who saw climate change as the 'most serious issue currently facing the world as a whole' was reduced to 61 per cent and 'poverty, hunger and lack of drinking water' was increased to 80 per cent. The EU average on the latter marginally decreased from the 2009 level (down to 64 per cent from 66 per cent), whilst concern for climate marginally increased (up to 51 per cent from 50 per cent). As argued in Chapter 3, research has demonstrated that environmental concern tends to decrease when set against certain materialist issues, like prices and employment (see Marquart-Pyatt, 2007). It therefore makes sense that concern about climate change decreases at a time when the Greek public faces an extreme deterioration of living standards. It is clear, then, that the professed concern about environmental issues in 2009 was very much conditioned by the aforementioned international and national factors rather than a rise in environmental awareness stimulated by the Olympics.

(iii) International treaty ratification

Like Australia, Greece was slow to ratify the Kyoto Protocol, albeit for rather different reasons. There was no outright justification based on arguments pertaining to the national interest, as in the Australian case. Greece was the last EU15 member state to ratify the protocol in 2002, within 24 hours of a 31 May deadline (Psaropoulos, 2008). It is important to note that the agreement allowed Greece to reduce its continuous increase of CO_2 emissions to 25 per cent on the 1990 baseline for the period 2008–2012. That was formulated in order for the country to make a fair contribution to the common EU objective of 8 per cent reduction for that period. Even though Kyoto sought a 12 per cent reduction by 2010, Greece secured an increase because it was seen as a developing country and as such the country's capacity to meet its targets has been heavily circumscribed by a vigorous developmental process with many polluting emissions (Elafros, 2007).

In order to secure its compliance, Greece formulated the National Plan for the Reduction of CFC Gas Emissions (2000–2010) in 2003. Nevertheless, Greece failed to fulfil its programmed commitments (8 per cent reduction) and was therefore excluded from the European emissions trading market until 2013. Greece is to participate in the mechanisms supported by articles 6, 12 and 17 of the Kyoto Protocol until it improves its capacity to fulfil the objective of 20 per cent reduction by 2020 and employs a reliable mechanism for measuring industrial emissions (see Bizios, 2011). In my earlier exploration on the subject (Karamichas, 2012a, p. 165), I argued that 'students of comparative European environmental policy may well not be surprised by the precarious nature of Greece's ability to meet the requirements of an agreement to which it has been a signatory. Greece, like its south European counterparts, has been notorious for its failure in applying EU environmental directives'.

There is considerable work on this area that I followed in my earlier treatment of the same issue (Karamichas, 2012a, ibid.). In the past I supported the notion that any dealing with this issue has to make at least a passing remark on works engaging with this lack of compliance with EU directives by southern European EU member states such as Greece: the 'Mediterranean Syndrome' (La Spina and Sciortino, 1993) and the 'Southern Problem' (Pridham and Cini, 1994). To further expand upon how these two perspectives explain non-compliance by pinpointing a range of socio-historical factors, their key components can be presented in the following way:

1. the lack of a civic culture, which would enable the development of cooperative behaviour among the citizenry; and mistrust toward state authorities, which inhibits compliance with the law;
2. ineffective and inefficient administrative structures, often characterized by a severe lack of technical expertise, which are intimately related to the permeation of patronage and clientelistic tendencies in the recruitment of officials; and
3. a party-dominated, reactive legislative process that impedes regulatory continuity.

It is quite difficult to deny these insights, especially in relation to both the issue that we are dealing with here and the current financial crisis. However, it is important to include some works that challenged these perspectives. The challenging perspectives offered by these works (Börzel, 2003, Koutalakis, 2004) stress that

> as relatively late entrants into the European Community, southern European member states have had to import policy and legal frameworks and adapt to the regulatory standards and administrative structures of northern European states, and thus face increasing implementation and compliance costs. Moreover, non-compliance can only be averted if the European Commission opens infringement proceedings – often after campaigning by domestic environmental movement organisations – thus raising the financial and reputational costs of non-compliance beyond the costs of adaptation and implementation.
>
> (Karamichas, 2012a, p. 165)

In a more recent examination of the ways in which the domestic institutional context in south European member-states has affected the relatively recent (since the early 1990s) EU preference for more cooperative policy schemes instead of hierarchical top-down approaches for improving compliance with four domestic environmental directives, the labelling of these countries as environmental laggards was challenged (Fernández et al., 2010). It is worth interjecting here with the fact that the originators of the Mediterranean Syndrome thesis, La Spina and Sciortino, were rather sceptical about the extent to which the Spanish case (in light of the fact that its administrative structures and legislative process play by the rules when dealing with serious social conflict) fits neatly with a 'disease' that could have been easily called, by their own admittance, the 'Italian Syndrome' (see Karamichas, 2007a). As such, there were always some exceptions within the overall categorization. The findings by Fernández and colleagues (2010) confirm that irrespective of the commonalities that these countries share in their historical, geographical, and cultural backgrounds 'as well as comparable levels of EU-national policy misfit at the time of accession [...] Spain ranks higher in the adoption of environmental governance mechanisms, with more regular systems of consultation, relatively dense and interconnected policy communities, particularly in the nature conservation domain, and greater institutional leadership and readiness to involve non-state actors in the technical aspects of policy' (Fernández et al., 2010, p. 572).

Greece was found to be at the opposite end with 'less structured coop-erative arrangements, irregular patterns of consultation and more importantly, a culture of administrative mistrust, impairing public-private cooperation' (ibid.). This was not always the case, though it was perhaps the most prevalent case in the post-dictatorship period. There were some pockets of change during the 1990s, like the mod-ernization attempts in the mid-1990s stimulated by both 'push' – EU policies – and 'pull' – environmental movements. Nevertheless, after a thorough account in relation to the four selected directives, the conclusion drawn is that 'despite considerable external stimuli from the EU, cooperative modes of policy-making remain weakly institu-tionalized. The domestic institutional context has not encouraged the emergence of cooperative arrangements on a regular basis. The weak coordination skills of state actors in pooling dispersed resources in terms of experience and expertise along with the weak capacities of the actors from civil society themselves to compensate for the State's poor cognitive and material resources have mediated EU pressures for change. However, the deepest impediment is the lack of institu-tional trust, which has conditioned relationships between state and non-state actors in pooling resources and sharing adaptation costs' (ibid., p. 571). The situation became more acutely problematic in the post-Games phase 'under George Souflias' environmentally-destruc-tive management of YPEHODE (the Ministry for the Environment, Planning and Public Works) during the 2004–2009 tenure of the conservative ND government of Kostas Karamanlis'. According to Psaropoulos (2008), '[u]nder Souflias [...] Greece has taken spectacu-lar steps backwards, even as Europe has moved forward. Souflias has never attended a European council of environment ministers. He was not only absent from December's round of UN climate talks in Bali, but failed to organise a representation under the deputy minister for the environment'. Indeed, Souflias's evident disdain for the environ-ment exasperated Stavros Dimas, a fellow New Democracy politician, during the latter's term as European Commissioner for the environ-ment (also 2004–2009)' (also cited in Karamichas 2012a, p. 165).

As has been indicated, the rise of Papandreou's PASOK alongside the development of an ambitious EM programme raised hopes that the country's extremely negative environmental standing would be corrected. One of the first actions taken was to disassociate the portfolio of the Ministry for the Environment from the planning responsibilities of YPEHODE, a perennial demand of Greek ENGOs who had witnessed environmental issues sidestepped by developmental and planning requirements. Similarly to the Australian (and British) case, the new Ministry incorporated the all-inclusive 'climate change' parameter in its overall portfolio. The Ministry for the Environment was renamed the Ministry for the Environment, Energy and Climate Change (YPEKA) and T. Birbili was appointed as its first minister. Birbili has a doctorate in environmental management from Imperial College, London, and for a period she was also a contributor to Papandreou's speeches. She was eccentric in her ministerial style (dressing casually, keeping minimal decorations in her ministerial office), and she was the frequent target of cheap attacks by some popular media commentators who often downplayed the positive changes that she initiated. On some of these initiatives she also attracted hostile commentary by fellow PASOK cadres for her plans to reduce the construction allowance in Natura areas. Media reports placed her at the top of the list of possible replacements in any ministerial reshuffles that were likely to take place in the extremely volatile climate that has remained.

That eventuality was realized in June 2011. Following the ratification in the Greek parliament of the interim programme, T. Birbili was replaced by a former minister of finance, G. Papaconstantinou, on 17 June 2011. Her replacement by a political figure with limited appeal among the Greek public, in light of his leading role in the most severe austerity measures and his opposition to some of her initiatives, signified a blow to the green development ambitions of Papandreou's PASOK. A member of the Ecological Society of Rodopi expressed the view that with Papaconstantinou at YPEKA the environment 'would be dealt with in accountancy terms' (Econews.gr, 2011a).

In a letter undersigned by 11 ENGOs, Papaconstantinou was called to 'ensure that the economic crisis would not be used as a pretext for the further devaluation of the environment in Greece with an aim to short-term and doubtful benefits' (Econews.gr, 2011b). With the speech that Papaconstantinou delivered at the 17 UNFCCC in Durban on 20 November 2011, he attempted to allay these fears:

> In a difficult economic situation for our country, we should not forget the future of our planet. The objectives and tools possessed by our country to confront climate change are not independent

from the effort to reset our economy and the provision of better living conditions to all citizens. Investments in Renewable Sources of Energy are new sources of income and places of work for the country, while they also strengthen our energy autonomy. Energy savings is a means of good individual household economy, it is however also a means for invigorating certain branches of our economy, such as material production, manufacturing and the construction industry. The reduction of CFC emissions by the industry upgrades our environment and quality of life, while it strengthens research, technological growth and at the end the competitiveness of the Greek economy. We move forward decisively towards achieving our climate change objectives that can offer vital prospects to our economy and green growth.

(Anon, 2011)

Papaconstantinou appears to be committed to the green growth path that has been advocated by the programmatic declarations of Papandreou's PASOK (see Pagoulatos, 2010). As far as the EM indicator of international treaty ratification is concerned, then, Greece appears to occupy an ambiguous place. Indeed, the country holds a more strenuous position as it is closely supervised by a troika of bailout lenders, which includes its EU partners; still, the exploitation of its green assets (solar and wind energy) for growth realization is something that fits well with their proposals. This does not, however, guarantee that earlier inhibitors would not operate again in the same negative ways. Most importantly, although this indicator has not been awarded a negative score, this does not necessarily indicate that hosting the Olympic Games played any direct role in this direction.

(iv) Protected natural site designation

Greece has 359 sites incorporated in the Natura 2000 European Ecological Network, and as a signatory to the EU Habitats directive, it holds a number of obligations in relation to the conservation of the habitats and species that occupy these sites (Dimopoulos et al., 2006, p. 175). Nevertheless, like Australia, Greece has been beleaguered by the annual occurrence of wildfires, which cause immense destruction in acreage. The death toll caused by the 2007 forest fires appeared to stand out as a turning point in Greek socio-political attitudes (see Karamichas, 2007). In an article on the 2007 forest fires, I pointed out that Greece lacked an official forest and land register, and this issue has sustained the practice of converting to building land the forest lands that are affected by fires. These lands were supposed to be protected for their rejuvenation. In interviews with ENGO representatives, members of the scientific community and state employees, Apostolopoulou (2009, p. 126) found that most of these people were intent on 'blaming the reluctance by the government to impose environmental conditions in economic policies':

> In addition, they supported that ambiguity is a standard institutional practice that limits culpability and essentially doesn't identify guilty parties and that way creates substantial obstacles in the application of nature protection laws. Ministry employees highlighted that these problems are worsened by changes of managers alongside changes of government or even ministers – something that leads to a piecemeal and partisan policy process.
>
> (Ibid., my translation)

This was the prevailing situation at YPEHODE that was described by Fernández and colleagues (2010). Nevertheless, when I tackled the issue of the non-existent land and forest register in my earlier treatment of that EM indicator, I noted that 'one of the first initiatives put in action by the new ministry of Environment, Energy and Climate Change (YPEKA) in December 2009 was a new legal framework for the rejuvenation of the burned forest lands of the Greater Athens area that prohibits any building in these areas before the composition and validation of a forest land register, finally passed by parliament in January 2010' (Karamichas, 2012a, p. 166). This development met a perennial demand of the Greek environmental movement. Since the replacement of T. Birbili, however, the forest and land register has been substantially altered and a range of her policy initiatives completely scrapped. I will provide a more detailed presentation of these changes in due course; for this section, I want to highlight that, with all the obstacles and challenges that T. Birbili faced during her 18-month term as YPEKA minister, she still managed the ratification of the first forest maps and that could be considered a great success (Lialios, 2011). Perhaps if this process had not been interrupted, it could have been argued that, with Birbili in YPEKA, Greece was finally on a path toward making a sincere commitment to the protection of its natural environment. Still, this argument may only further substantiate the earlier assertion that the intricacies of national politics play a principal role in the evolution of EM capacity. Another idea is that Papaconstantinou has fully embraced the green development perspective and is continuing along the EM path without exhibiting the sensitivities of his predecessor. Overall, there is still too much ambiguity to award with confidence either a positive or negative score on this indicator. In any case, this score has little to do with Athens hosting the XXVIII Olympiad.

(v) Environmental Impact Assessments

As indicated earlier, EIAs had to be performed for all Olympic-related projects. However, that initiative appears to have had little post-Olympic impact. In fact, inherent problems in the EIAs were amplified in this case by an absence of attempts to overhaul poor administrative practices. Two studies on EIAs in Greece, one conducted before and the other after the Games, reached similar conclusions. Androulidakis and Karakassis (2006) evaluated the quality of environmental impact studies produced in Greece between 1993 and 2003 and found that they performed rather poorly with respect to most indicators and that there was little evidence of improvement. Botetzagias (2008a) conducted nationwide research and reached similar conclusions, adding that the lack of specialized staff and material shortages in many local authorities severely compromised the effectiveness of the EIAs. The importance given to the exploitation of renewable energy sources as part of the 'green growth and development' agenda and 'the need to address the country's unexplored potential in renewable energy development' (Pagoulatos, 2010, p. 6) by the Papandreou government was bound to attract the investment interests of key corporate actors. The location of certain renewable energy facilities, such as wind farms, has been contentious; challengers cite the complete absence or violation of EIA procedures and findings. Thus, we have a case where environmentally contentious projects are simultaneously connected to green benefits and growth. Therefore, EIAs are performed in a swift way that is favourable to corporate interests. In the context of the severe crisis that Greece has been facing, we are confronted with the same problem areas that have characterized other EIA processes – inherent ambiguity and pressures for project completion – but with the twist of hiding locally unwanted/undesirable land use (LULU) under the guise of green development. The EIA indicator has again achieved a low post-Olympics score.

(vi) ENGO participation

A budding interest in environmental issues in the mid-1990s appeared to expand in the early 2000s through the PASOK government's appointment of a number of environmental activists to government and advisory positions. As a case in point, the former director of Greenpeace-Greece, Elias Efthymiopoulos, was appointed as a junior YPEHODE minister; Efthymiopoulos projected the concept of EM when questioned about the compatibility of environmentalism with the modernizing project that the government was putting into action (see Karamichas, 2008). 'With hindsight,' however, 'it is obvious that that collaboration did not produce the expected results on the environmental front, and the proclaimed modernization project of the 2000–2004 administration produced only limited results, if any' (Karamichas, 2012a, p. 168). According to Papadopoulos and Liarikos (2007, p. 302), there was a partial institutionalization of ENGOs through their participation in a number of steering committees; however, this was 'largely undefined and based on informal interaction'. As I have highlighted, the environment was further downgraded during this period. In my earlier treatment of this item, I claimed that '[u]nder the current PASOK government, with its highly ambitious green plans, it appears that there is a *de facto* openness of YPEKA in collaborating with ENGOs and accepting their input into policy-making' (Karamichas, ibid. p. 168). Indeed, ten ENGOs proclaimed that the establishment of a green fund under Birbili 'was a positive development for environmental policy in Greece, as it raised hopes that the environment would finally receive what has been collected for its protection. It also raised hopes that the distribution of funds to projects like urban and suburban restoration [...], energy upgrades of the building reserve and the re-establishment of downgraded biotopes would have given the essential impulse to "green" economic activities that are still very limited in Greece' (Odysseas, 2011; my translation). Nevertheless, with changes brought in by Papaconstantinou, '95 per cent of the funds collected for the Green Fund for the regulations of semi-open spaces and the legalization of buildings lacking planning permission, through emissions trading and income from fines for environmental crimes, will be incorporated to the general government budget' (Odysseas, ibid.).

Discussion

In my earlier study of the post-Olympics EM capacity of Australia and Greece (Karamichas, 2012a, p. 169), I concluded that 'not only has there been no noticeable post-event "environmental improvement" [in either country], but there has also been an evident continuation of practices leading to a progressive worsening of the environmental status of both nations'. In that work, one indicator – 'level of environmental consciousness' – achieved a positive score. With the economic downturn, there is a notable decrease in the numbers expressing high levels of environmental concern, albeit for slightly different reasons. In both countries, the environment is downgraded as an issue of importance when immediate 'bread and butter' issues are more visible. A counter-argument could perhaps support the idea that the 'level of environmental consciousness' may have been even lower had it not been for the Games; in the Greek case that view has been examined using the Eurobarometer survey, which ensures a cross-national European comparative perspective that clearly demonstrates that the economic strain has caused an overall decrease in environmental consciousness across Europe. It also reveals that Greeks tended to exhibit extremely high levels of environmental consciousness during the early 2000s. In Australia, the decrease had more to do with a return to 'normality' after the unprecedented rise recorded in 2007. Table 7.1 shows that this was not the only indicator that attracted a negative post-Olympics score.

Table 7.1   EM in the post-event phase – Australia and Greece

	Sydney 2000	Athens 2004
Annual level of CO_2 emissions	☺	☺
Level of environmental consciousness	☹	☹
Ratification of international agreements	☺	☺
Designation of Sites for Protection	☺	☺
Implementation of EIA procedures	☺	☺
ENGO participation in decision-making processes	☺	☹

'ENGO participation in decision-making processes' in the Greek case also attracted a negative score, after the heralded opening to their viewpoints that was manifested by Birbili was downgraded to the realm of uncertainty under Papaconstantinou. A similar uncertainty also characterizes the prospects of that indicator in the Australian case and all four of the remaining indicators in both cases. The overall conclusion points to a lack of correspondence between the EM aspirations of Olympic Games hosting and the actual capacity for EM by these two host nations.

Beijing 2008

This section examines Beijing, an Olympic host city as well as the capital of an autocratic regime, the People's Republic of China (PRC). Notably, the PRC 'has become the most powerful engine for global economic growth' (Watts, 2011b, p. 23). Before exploring China's EM capacity after hosting Beijing 2008, however, it is necessary to dedicate some time to the regime itself. This will facilitate a better appreciation of the rationale that underpins the regime and the impact that a lack of liberal democratic frameworks may have on the post-Olympic EM of the country.

This discussion must commence by offering some insights about the host city. After all, prospective host cities tend to see the Games as an opportunity for the restructuring and regeneration of the city and/or the area of the city in which Olympic facilities are to be located (Home Bush Bay in Sydney, a substantial part of Athens, Stratford in London).

Beijing, which is the capital of PRC, has a population of 15.3 million, based on 2005 data provided by the Beijing Municipal Bureau of Statistics (BMBS) (Xie, 2009, p. 86), yet 'the permanent registered population measured 11.08 million'. According to Broudehoux (2012, p. 60), 'Beijing is estimated to have a migrant population of about four million, mainly employed in the construction, manufacturing, and service industries, within a total municipal population of around 17.5 million'.

Leaving aside the different measurement of the total population of Beijing, it's important not to lose sight of the fact that this migrant population, along with their contribution in the preparations for hosting the Games, was conveniently forced to disappear for the duration of the Games – along with a range of other unwanted elements (beggars, street children, the homeless, etc.) (ibid., p. 63). This is not an unusual occurrence in Olympic host cities. In fact, the work by Chris Olds (1998) and Centre On Housing Rights and Evictions (COHRE) (2007) on evictions of undesirable communities in preparation to host the Olympics or other mega-events, including beauty pageants, has demonstrated that liberal democracies are not immune to this phenomenon. Still, though, there are a number of issues in the case of Beijing that need to be teased out in order to fully explore China's relationship to both SD and EM. It is worth continuing along the lines of the treatment of migrant communities in Beijing. The following can be found in the 2008 report by COHRE:

> According to a number of people interviewed, the Chinese government, public, and media rarely view housing issues – particularly the government's role in creating an enabling environment for the provision of adequate housing – from a human rights framework. People rarely raise concerns about violations of their human rights, but rather state their concern in their own language.
>
> (COHRE, 2008, p. 16)

The latter finding puts under immediate scrutiny the BMBS boast that '[p]eople of all China's 56 ethnic groups are found in Beijing. The vast majority of the population belongs to the Han ethnic group [and] people of the Hui, Man and Mongolian ethnic groups number more than 10,000 separately'. The impression provided here is that Beijing is an ethnically homogenous society built of mainly Han residents, which nevertheless has room for other ethnic communities. After all, the preamble of the Constitution defines the PRC as a 'unitary multinational state created jointly by the people of its nationalities' (Beyer, 2006a, p. 188). Interestingly enough, during the opening ceremony of Beijing 2008, an attempt was made to demonstrate the unity of the country's 56 ethnic groups. Children dressed in costumes associated with these ethnic groups accompanied soldiers carrying the national flag. Leaving aside the allegations that the children were all from the Han Chinese majority, this parade was 'a sign of how sensitive ethnic relations in China are' (see Spencer, 2008). Moreover, in a direct link to the main theme explored here, it is important to note that the 'Green Olympics' concept was reflected in the five Olympic mascots. Four were animals representing natural elements. One of them, 'Yingying – a Tibetan endemic protected antelope showcased Beijing's commitment to Green Olympics and "Grass-covered Ground" idea' (UNEP, 2009, p. 16). However, for the Tibetan Women's Association, China's use of the Tibetan antelope as one of the Olympic mascots was an 'attempt to demonstrate the "unity of the nationalities in China" and to legitimize China's occupation of Tibet [...] The Tibetan antelope is not a Chinese symbol but it is being used as a Chinese propaganda to convince the world that Tibet is part of China' (TWA, n.d.). In fact, Tibet was an issue that stimulated waves of anti-Beijing protest activity around the world. These protest activities took advantage of the opportunity afforded by the global torch relay – a legacy of Athens 2004 – across 26 countries and all five continents (Payne, 2006) to promote the Tibet case (see Renou, 2012).

Another crucial aspect is that the Tibet issue, along with a range of other cases that constitute human rights violations perpetrated by the regime, can easily erase with one swoop any claims 'to sustainability in a wider definition' (Mol, 2010, p. 512). For that reason, I have adopted like Mol a more restricted 'environmental definition of sustainability'. I return to this issue in the final chapter, where I discuss the implications of such a strategy on the actual value of EM and the extent of its connection to SD. Meanwhile, by continuing along the same lines, we can bring to mind the fact that one month before a recommendation was made to invite China to join the group of eight of the world's leading nations (G8) in Savannah, Georgia (June 2004), 'then Prime Minister Tony Blair had met with President Wen Jiabao to discuss issues of democracy and human rights in relation to Hong Kong and Tibet [...] [During that visit, the Chinese President] was confronted by protesters from the Falun Gong Movement and the Free Tibet Campaign. The protesters at the time viewed Wen Jiabao as a part of the new generation of Chinese leaders and were reported to be remaining optimistic about the leadership and the issue of China's 54-year occupation of Tibet' (Jarvie et al., 2008, p. 138).

By recounting these two interconnected developments, Jarvie et al. (ibid.) wish to 'reflect the challenge that is China in the twenty-first century':

In China privatization is occurring before democratization, with China moving toward a closer relationship with globalization on its own terms – in other words, what has been referred to [...] as creeping toward capitalism with Chinese characteristics. Social, economic and cultural changes have been afoot in China following the perceived failure of Mao's egalitarian socialism in which an understanding of China's own approach to consumerism, Confucianism and communism is only part of the guidebook to making sense of China today.

With this in mind, I include the following interjection to facilitate a better understanding of China's place in the context of modernity and modernization.

China and modernity/modernization

For Wang Hui, a leading contemporary Chinese intellectual 'modernity or "early modernity" [...] is an open possibility rather than a structural project. The only thing we know about its meaning is that it involves the emergence of new pathways, not the replication of any version of modernity confected in the West' (Zhang, 2010, p. 54):

> For him, it was anachronistic to picture socialist China as if it were a feudal autocracy, and naive to regard the market as an escape hatch from repression by the state. In his eyes, the Chinese Revolution remained an important source of criticism for the current society.
>
> (ibid., p. 52)

Wang Hui has drawn parallels between the student movement of Tiananmen Square in 1989 and the May Fourth movement of 1919 (which signified the upsurge of Chinese nationalism), as they shared some 'self-defeating elements'. Both movements were driven by students who demanded democratization. Moreover, 'the students in Tiananmen Square had shared much the same uncritical faith in science as the party leaders they opposed. In the 1990s, the cult of science was then adjusted to support the shift towards marketization and privatization of the post-Tiananmen period' (ibid., pp. 52–3). Although Wang Hui vehemently opposes the view put forward by Weber at the time of the May Fourth movement 'that at this stage pre-Qin China had exhibited a kind of political rationality comparable to that of early modern Europe', he does not dispute that 'Post-Qin

China was very dynamic, manifesting a different kind of rationality, unknown to Weber' (ibid., p. 74). Whether that was modernity European-style or purely Chinese, the truth is that China has not been singularly opposed to the modernizing process. China's initiation to modernization can be traced back to Den Xiaoping's elevation as leader of the Chinese Communist Party (CCP) in 1977 and his determination 'to take the country in a new direction, opening up China to the world and liberalizing its economic and foreign policies' (Xu, 2008, p. 197). In December 1978, the CCP announced the Four Modernizations programme in agriculture, industry, defence, and science and technology (Li et al., 2007, p. 209). This programme has been seen as the opening of China to the world. For those who might have viewed this development as an abandonment of Communist principles for pragmatism, it is worth remembering 'Deng Xiaoping's cat theory which implies that it does not matter whether the cat is black or white or red – as long as it catches the mice it is a good cat' (Jarvie et al., 2008, p. 95). Along these lines, 'For many Chinese people it does not matter whether the cat is socialism with Chinese characteristics of capitalism with Confucian colours – as long as it works for China's development it is likely to be viewed as a goodism' (ibid.). As Cook (2007, pp. 287–8) has put it:

> The Dengist model took China, probably irrevocably, down the capitalist road – albeit under strong direction from the Chinese state. It was based on an open door for foreign direct investment (FDI), with the objective of modernizing China's agriculture, industry, defence, science and technology. Under the market reforms unleashed by Deng and his successors, China's 'Gold Coast' has opened up to global connections and China is now in rapid transition from being a closed, poverty-stricken rural society towards being open, wealthy and, for many, urban. In brief, the Chinese state sets the preconditions for investment to enter, the local state (at province, city or town level) provides the infrastructure, and foreign companies provide the necessary investment through which China's resources of land and labour can be fully exploited.

Although 'China's desire to host the Olympic Games is nearly as old as the Olympic Games themselves' (Brownell, 2006, p. 52), Olympic Games hosting should be seen in the context of the Dengist plan 'through reform and opening up' of which 'China's sports policy in general and its relation with the Olympic Movement were part of' (Xu, 2006, p. 93). As such, the first Chinese Olympic Games candidature in 1993 was primarily aimed to showcase PRC's adaptability to the market reforms of the 1980s. Memories of the violent crackdown in Tiananmen Square also infused that aim with the need to 'reengage with the world community' (Worden, 2008, p. 25). When Deng Xiaoping announced in 1990 that China was applying to host the Games, officials devised a slogan, splashed in English on billboards and walls in Beijing, "A More Open China Awaits the 2000 Games" (ibid., pp. 25–6). However, in the words of a Chinese journalist 'when the application was made in 1993, the sounds of gunshots in Beijing were still ringing in people's ears' (ibid.).

Beijing narrowly lost its bid to host the Games in 1993 by two votes to Sydney (see Close et al., 2007, pp. 88–91). Human rights were 'cited as a major reason' (Cook, 2007, p. 286). Another important factor that contributed to Sydney's success was the environment. Although the environment did not play a significant part in the relevant Manual for Candidate Cities (MCC) (see Cashman, 2009), the Sydney bid stood out because of its commitment to 'use the Olympics as a vehicle for best practice solutions to address' a range of environmental issues (Zhang, 2008, p. 5).

Because of Sydney's green promise, the promotion of SD became an indispensable component in Beijing's bid in 2001 to host the 2008 Games (Loh, 2008, p. 240; Mol, 2010, p. 517; Mol and Zhang, 2012, p. 133). There was though an additional factor that exemplifies the development of the environmental impact of Olympic Games hosting beyond the landmark first 'Green Olympics' of Sydney 2000. Before the specificities of this development are revealed, it is necessary to interject commentary related to aspects of Olympic Games hosting that also played an important role.

A couple of commonalities with the Greek case can be detected here. In both cases, two bids had to be submitted and national pride was an important and stimulating factor. Indeed, the latter factor appeared to have played a major role behind the immense public support for hosting the Beijing Games. 'The Bid Committee claimed 95% public support for hosting the Games and the IOC poll showed 96% support in Beijing and other urban areas' (IOC, 2000, p. 60). Interestingly, that was the highest popular support given to any of the bidders for the 2008 Games. According to Close et al. (2007, p. 113) 'the greatest popular support was given to those candidate cities which were least developed while, none the less, exhibiting the greatest economic growth rates, at least as indicated by increases in gross domestic product (GDP)':

> the announcement in June 2001 that Beijing would host the Olympics in 2008 was greeted by mass rejoicing. Even Shanghai, often cast as a rival of Beijing, witnessed warm celebrations when the result was announced [...] The announcement provided a sense of vindication of China's improved standing in the world. When the People's Republic of China (PRC) was founded [...] Mao Zedong said in his address to the new nation that China had stood up and would never be humiliated again. After many years of effort, of marked successes and notable failures, the opportunity to host the 2008 Olympics was proof that China had not only stood up, but that it was no longer a pariah state and was ready to take its rightful place as one of the leading countries on earth.
>
> (Cook, 2007, p. 286)

According to Brownell (2006, p. 52), '[w]hen Beijing won the bid and it was shown live on Chinese television, tens of thousands of people took to the streets to celebrate and the optimism about the Olympic Games and the future of China [was becoming] more fervent as the Games [drew] near'.

When it comes to the environmental issue, Beijing stands diametrically opposed to Athens and Sydney. The reason for this lies in the fact that, in the award of the Games to Sydney, 'the environment hardly played any role, either in (criteria for) the bidding document prepared for IOC decision-making, or in the discussion and decision-making process itself. This aspect was notably different in Beijing's bidding in 2001 for the 2008 Games. Beijing used Sydney as the example for budgeting their Olympics, including the latter's significant, path breaking attention to the environment and green Olympics' (Mol and Zhang, 2012, p. 133). Moreover, although not a formal requirement, when Beijing 2008 submitted its bid, it applied an OGI study as agreed upon by the IOC in 2001 and as such was the first Olympic host city to produce such a study. In effect, what we can clearly see by examining the 'pre-event' and 'event' phases of the XXIX Olympiad is a paradigmatic case of a meticulously organized sport mega-event in relation to the environmental factor.

The environment and the bid

The Beijing bid stipulated from the outset its environmental commitment:

> 'Green Olympics' is one of the three key themes for the Beijing 2008 Olympic Games. This theme owes its prominence to the Chinese philosophy dating back 2200 years, which recognises a connection between sustainable use of the environment and human existence. The 'Green Olympics' is integral to the planning and staging of this great event.
>
> With the Olympic ideal as the major catalyst, 20 major projects costing US\$ 12.2 billion aimed at improving the environment shall be completed by the year 2007, achieving the objectives set forth in the city's Master Plan for Development three years ahead of Beijing's schedule. Beijing promises to provide a clean environment for the athletes by 2008.
>
> (BOCOG, 2008, p. 49)

The IOC (2000, p. 62) produced the following commentary on the environmental component of the Beijing bid:

> Beijing currently faces a number of environmental pressures and issues, particularly air pollution. However, it has an ambitious set of plans and actions designed and comprehensive enough to greatly improve overall environmental conditions. These plans and actions will require a significant effort and financial investment. The result would be a major environmental legacy for Beijing from the Olympic Games, which includes increased environmental awareness among the population.

The decision to hold the 2008 Games in Beijing brought specific environmental remediation measures into focus, including 'significant environmental cleanup activities' (Brajer and Mead, 2003, p. 240).

In relation to this United Nations Environmental Programme (UNEP, 2009, p. 13) made the following comments:

> Beijing had a very ambitious programme to offer a 'Green Olympics' to the world. Several targets of the Beijing 'Environmental Master Plan' (an environmental protection programme developed by the Municipal Government for the period 1996–2015, funded by the World Bank) were integrated into the bid with accelerated deadlines. Some targets, originally scheduled to be achieved in 2010, were moved forward to 2008, the year of the Games.

After all, 'convincing the IOC that Beijing could clean up its environment was critical to winning the prize to host the Olympics' (Loh, 2008, p. 240). In the end, environmental protection constituted 60.05 per cent of the total expenditure incurred in all investments related to the Beijing Games (Brunet and Xinwen, 2009, p. 166).

The following illustrative factors have been put forward by Jarvie et al. (2008, p. 126) to demonstrate how the Beijing Olympic Games Bidding Committee (BOBICO) shaped the bid in relation to the environmental component:

- Beijing's air pollution was thought to be one of the factors in its loss to Sydney in 2000.
- Since 1998, UD$ 15 billion per year have been invested in transport, communication and environmental improvement.
- In October 2000, the city banned the setting-up of barbeque stalls in an attempt to reduce air pollutants.
- In November 2000, the city with the World Bank launched a US$ 1.25 billion initiative to help Beijing approach World Health Organization (WHO) clean air standards for cities by 2006, and 22 measures to reduce smog were introduced, including the closing of local steel mills by 2002.
- Tianjin, China's third largest city, has launched six major projects aimed at Tianjin (a co-Olympic host city) becoming part of the Beijing–Tianjin Ecological Zone.
- In April 2001, a total of 2008 trees were planted in a park in Beijing to express support for a 'Green Olympics'.

According to Brajer and Mead (2003, p. 240), many of these efforts represented 'some sort of long-term or permanent cleanup'. In particular, notable actions taken since 1998 include the conversion from coal-based to natural gas heating and cooking, and the implementation of factory and automobile pollution emission standards (ibid.). By 2000, the Beijing Environmental Protection Bureau (EPB) was able to report that these cleanup efforts resulted in a 44 per cent reduction in SO_2 pollution and two-year decreases of 13.7 per cent in total suspended particulates (TSPs) and 14.5 per cent NO_2 (ibid.).

Pre-event actions

In 2002, as part of fulfilling the environmental pledges contained in the Olympics bid, Beijing released the Olympic Action Plan, which with the ambition of making Beijing an 'ecological city' developed a five-year plan that pledged to deliver a number of improvements in order to mitigate the CO_2 emissions and polluting effluents produced in the city. By 2008, many of the pledges had been carried out or were in place (Loh, 2008, p. 214). That was facilitated by laying out 'an ambitions timeline dividing the renaming period until the opening of the Games into different operational stages' (Beyer, 2006b, p. 429).

In addition, in 2005 Beijing Organizing Committee for the Olympic Games (BOCOG) signed a Memorandum of Understanding with UNEP. 'The agreement laid the foundation for UNEP to support BOCOG on the greening of the Games. As part of the agreement, UNEP conducted a review and published a report on the environmental performance of BOCOG in October 2007. The report reviewed the environmental efforts of the Committee in meeting its commitments to organize a "Green Games" and concluded that Beijing was on track for achieving the promises it laid down on its 2000 bid' (2009, p. 13). Still, in 2006, the international community was still voicing concerns about the possible failure of China to put into action environmental policies. The fact, that it took the government 10 days to publicly acknowledge an 80-kilometre-long slick of highly toxic benzene, which had contaminated the water supplies for Harbin, one of China's biggest cities, did not help to allay these fears (Beyer, 2006a, pp. 186–7).

Among the package of measures announced in 2007, we find the following:

- reducing the number of vehicles on the road during the Games;
- eliminating dust from construction sites and cutting industrial emissions;
- an emergency plan if pollution is high during the Games that includes 'shooting chemicals into the air to seed rain to damp down pollution and hopefully clear the skies'; and
- 'the Beijing Sustainable Development Plan with 'twenty key projects and various anti-pollution measures, such as preventing coal burning by retrofitting power plants with scrubbers'. (Loh, 2008, pp. 239–240)

All of these measures were implemented with immense precision and efficacy. Mol (2010, pp. 518–9) has offered a good account: 'The environment permeated all Olympic processes – design and construction, refurbishment, marketing, procurement, logistics, accommodation, transport, office work, publicity and operational affairs' (ibid.). Indeed, Beijing was able to boast that most of the plans had been auctioned by 2007 and these far beyond existing requirements and expectations. For instance, all projects had to pass EIAs 'even though they were often not legally necessary, and 14 new wastewater treatment facilities were installed. Five new public metro lines were laid especially for the Olympics, thus doubling the mileage to more than 200 kilometres. Acceleration in leapfrogging vehicle emission standards was achieved, to Euro IV standards in 2007' (ibid.).

Nevertheless, closer to the Games it became apparent that these measures had not been effective in relation to the 'most contested issue in the run-up to the Olympics' – air quality (ibid.). 'Beijing has the dubious distinction of competing with Mexico City for the "honour" of the world's most polluted capital' (Beyer, 2006a, p. 193). UNEP (2009, p. 11) points out that this major issue of concern 'was dealt with via a mixture of forward-looking planning measures backed by short term ones such as controls on private and public vehicles'. Far more than this, the aforementioned emergency plan had to be put in action. Mol (ibid., p. 526) relates an example of the investment made by China in presenting its new face to the world:

> From 20 July 2008 onwards, a driving ban was imposed on half of the three million registered cars as an emergency measures. In addition, most construction in Beijing stopped. Over 150 companies in Beijing and in neighbouring province of Hebei – and Shanxi, Inner Mongolia and Shandong provinces – closed temporarily from two weeks before the Olympics started until the end of the Paralympics. A tropical storm that hit southeastern China on 28 July brought strong wind, some rain, lower temperatures and cleared the skies just before the Olympics started, helped a little by launching more than 1000 rockets with artificial weather modification and control chemicals.

On 8 August 2008, a date that auspiciously contained the Chinese lucky number of eight three times, the Games opened with a spectacular ceremony at the Bird's Nest Stadium in Beijing. Having passed the opening ceremony with flying colours, however, didn't mean that the Games were immune from pollution-related impacts on athletic performance. Although 'matters could have been worse had there not been' the above measures and weather change, 'researchers found that four fifths of the time athletes were exposed to levels of coarse particular matter higher than considered safe by the international health watchdog, the World Health Organization' (Jamieson, 2009). This fact does not undermine the result, which was that an important opening to environmental sustainability in both the institutional and civil society had taken place amidst preparations to host the Games. The following section tests the extent to which evidence points to improvements in the EM capacity of China three years after hosting the Olympiad.

EM in the post-event phase

(i) Average annual CO_2 emissions

Although it was expected that China would achieve the dubious honour of becoming 'the world's largest greenhouse-gas polluter in 2020' (Beyer, 2006a, p. 188), it managed to 'become the world's largest greenhouse gas (GHG) emitter' in 2006 'and is now contributing

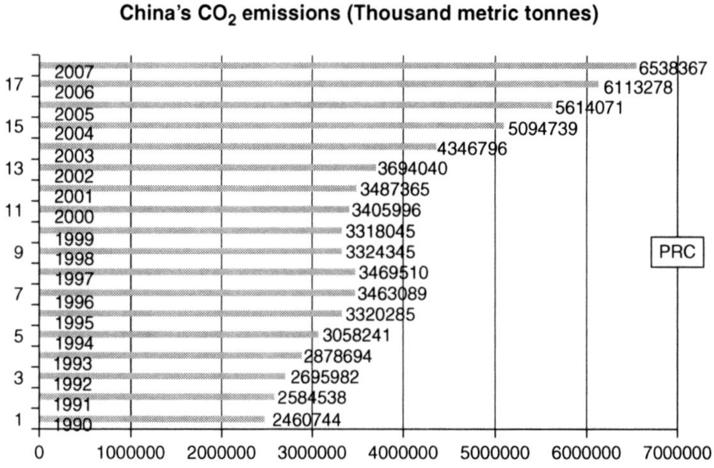

China's CO_2 emissions (Thousand metric tonnes)

Year	Value
2007	6538367
2006	6113278
2005	5614071
2004	5094739
2003	4346796
2002	3694040
2001	3487365
2000	3405996
1999	3318045
1998	3324345
1997	3469510
1996	3463089
1995	3320285
1994	3058241
1993	2878694
1992	2695982
1991	2584538
1990	2460744

PRC

Figure 7.5 China's CO_2 emissions (kt)
Source: UNSD (2010).

to raising CO_2 concentration levels' (Lin et al., 2009, p. 341) by being the number one energy user and arguably the most polluted nation on earth. 'The International Energy Agency noted that Europe's plan to extend 1990–2010 carbon dioxide cuts from 20% to 30% would equal only two weeks of China's emissions' (Watts, 2011a, p. 29).

Figure 7.5 shows that from 2001 to 2007 – six years after Beijing was awarded the Games – CO_2 emissions increased by 87.5 per cent. Measures taken in Beijing in 2007 to reduce pollution levels before the Olympics have already been accounted for. Naturally these measures were not potent enough to reduce the country's GHG emissions, especially at the time when the country continued to produce the same industrial output. At a time when China was still behind the United States in its standing as a GHG emitter, Richerzhagen and Scholz (2008, p. 314) suggested that 'GHG emissions are expected to increase in the future, because almost 90% of China's CO_2 are due to energy use [...] Estimates indicate that China's total primary energy consumption will more than double during 2000–20'. The available evidence points to a worrying increase in energy intensity since 2002 and a reversal in general energy intensity from 1978 to 2006, which showed a notable decline 'by 68% overall and by 4% per annum' (Lin et al., 2009, p. 344).

Marks (2010, p. 985) suggests that the numerous critics in the 'climate change community' who

> have lambasted Chinese leaders for ruling a country that is the world's largest emitter and for not doing enough to curb the country's emissions should consider that one-third of China's CO_2 emissions result from exports headed to developed countries and that the country has only emitted 7% of cumulative emissions between 1850 and 2000 (whereas the US and EU each have accounted for almost 30%). Further, the community could better appreciate the policy dilemmas that these leaders face: they cannot sacrifice economic growth without weakening their legitimacy in the short-term, and yet their future legitimacy could be jeopardized if they keep practicing business-as-usual.

Economy (2007, p. 33) offers some important illustrative points in relation to this situation. In her view, 'the more pressing issue for China's leaders is how to ensure that the extraordinary level of foreign direct investment can be channelled in environmentally constructive rather than destructive forms'. She illustrates this point by using Lee and So's argument (1999, p. 4) that 'many multinationals transfer "substandard industrial plants and hazardous production processes" to Asia in order to avoid the health and pollution standards in their home countries'. In what appears to be a full subscription to Kuznets, she concludes that others 'have suggested that economic development and integration into the international economy can positively affect environmental protection by optimizing efficiency of resource use, upward harmonization of environmental product and process standard, and increasing the standard of living, thus raising the demand for environmental protection' (ibid.).

Interestingly enough, in 2010, 'China invested $34bn in clean technology, compared to $18bn by the US. The contrast – which shocked many in Washington – is partly explained by different political systems, vested interests, and stages of development. While the US is dominated by big oil and big money, China is run by big hydro and big brother – a dictatorship of engineers' (Watts, 2011, p. 29). Moreover, in its twelfth five-year plan (March 2011), which was 'hailed as the greenest strategy document in the country's history', the PRC outlined a number of initiatives to green the Chinese economy, with some of them targeting a reduction of GHG emissions, such as declaring energy efficiency and environmental services as 'priority industries', introducing a carbon intensity target, which set the ratio of GHG emissions relative to GDP and a mandatory carbon trading system on a regional level (ibid.).

Economic growth and present/future legitimacy have also permeated the overall score of other EM indicators. At this stage, it appears that, in light of the 2011 five-year plan, China is fully committed to reducing its GHG emissions and is positive on the first EM indicator. Three years after hosting the Games, China has not yet reached the plateau stage in the Kuznets curve (see Chapter 5) to see a manifest reduction of its CO_2 emissions, but it is setting the preconditions to achieve that goal.

(ii) Environmental consciousness levels

In 2006, Beyer (2006a, p. 186) suggested that, although '[d]irect transfers of Western solutions are unlikely to succeed in contemporary China [and assuming that economic growth remains the priority, there is a need for the country to] effectively enforce its legal regime without compromising environmental concerns, public awareness of environmental issues, as well as the possibility of public participation by citizens and non-governmental organizations (NGOs) that provide expertise and accelerate the implementation process, are important factors'. Is it true that environmental consciousness has increased among the population after the 2008 Games?

According to Lo (2010, p. 1016), '[c]ompared with other countries, the level of concern expressed by the Chinese public is in general modest. Nevertheless in a report [published in 2007] summarising several global surveys [...] China has a greater proportion of citizens who have heard of global warming than do other developing countries. However, few Chinese individuals rated global warming as a "very serious" problem and few strongly believed it would pose a threat to them – fewer than in many other developing countries. Along with Americans, Chinese were those least likely to worry about climate change'. Interestingly enough, Giddens (2009, p. 103) found that a 'cross-cultural study of nine developed and developing countries indicated that about 60 per cent of people interviewed about climate change in China, India, Mexico and Brazil felt a "high level of concern", in contrast to 22 per cent in UK and Germany'.

A 2009 global public opinion poll, conducted by the University of Maryland's Programme of International Policy Attitudes, found that in 'China there was an overwhelming support – 94% – for the government to keep climate change on the front burner' (Goldenberg, 2009). In recent work, Marks (2010, p. 983) claimed that only 10 per cent ranked 'environmental concerns as the nation's number one issue', but it did rank 'as the fourth-highest concern among the country'. Yet another poll 'revealed that 62% of Chinese believe their country should reduce emissions by at least as much as other countries' (ibid.). It is envisaged that 'this increased perception could lead to increased pressure on officials to curb pollution' (ibid.).

Can we claim that Olympic Games hosting has contributed to increasing the levels of environmental consciousness among the Chinese public? According to UNEP's report on the Games,

> [w]ith such great citizen participation and media attention on all Games-related activities, the effects of activities on the environment, the health of the population and of the surrounding ecosystems became common-speak. For the average citizen of Beijing, environmental issues that might have, in the past, been taken for granted became major concerns. People began to appreciate these issues on their quality of life.
>
> Air quality, more than any other environmental concern, dominated media reports on the Games, mainly because of the direct impact on the health and performance of the athletes. Government officials, athletes, sport officials and fans, were all drawn into the debate on whether the air quality of Beijing could be safe for Olympic sport, particularly for endurance events. It created a media-frenzy in which journalists and media houses debated contrasting viewpoints and led to a general probe on the validity of official data released by the Beijing Environment Protection Bureau (EPB). Several other issues including energy, water and public transport also attracted media coverage, and thereby helped in kickstarting another debate and raised public awareness on these issues.
>
> (UNEP, 2009, p. 18).

Like the Greek case, hosting the Games seems to have been a significant contributor in raising levels of environmental concern among the Chinese public. Furthermore, it is notable that 'environmental consciousness' was not the only EM indicator with a positive score.

(iii) International treaty ratification

According to Beyer (2006a, pp. 185–6) 'China's environmental legal framework is rather young and started almost from scratch in the early 1970s after the country's attendance at the Stockholm Conference on Human Environment. In the subsequent era of China's open policy and reform, initiated by Deng Xiaoping, the environmental regime began to develop, coinciding with a growing awareness of environmental issues'. Nevertheless, this was not followed up by adequate enforcement measures and was very much conditioned by the decentralizing organizational character of Chinese policy administration and the extent to which economic growth was taking precedent.

China did not shun the international agreement aiming at climate change mitigation. In fact, 'China ratified the United Nations Framework Convention on Climate Change and the Kyoto Protocol in 1992 and 2002, respectively. As a non-Annex I developing country it hasn't been subject to emission limits. Although it doesn't lack aggressive mitigation efforts, there is a strong reluctance to curb skyrocketing emission levels by adhering to mandatory limits' (Lo, 2010, p. 1012). That was clearly illustrated in the December 2009 climate change talks in Copenhagen. As Brahm notes (2010, p. 40), China found itself on the defensive – and reacted with predictable intransigence. 'China, along with India, argued that the US should bear responsibility for historic CO_2 emissions and take the lead in reducing them'. Nevertheless, although 'China is unlikely to commit to any targets yet and many of its critics demand that China do more, Chinese leaders, wanting to make sure that climate change will not impede the country's development, have made a high-level policy commitment to tackle the problem. As a result, the country has steadily increased its use of renewable energy and percentage of forested areas, and has continued to curb the birth rate of its population' (Marks, 2010, p. 972).

In addition, 'after Copenhagen, China pledged to reduce its per unit of economic growth CO_2 emissions from 2005 levels by 45% by 2020' (Brahm, ibid.). After all, 'Chinese leaders are concerned about climate change because, as the Ministry of Science and Technology declared, "global climate has an impact on the nation's ability to develop further". China has already started to encounter problems from climate change, such as Tian Shan Mountain's loss of 20 million cubic meters of ice and the drying up of 17.5% of the lakes at the source of the Yangtze. As the level of emissions grows, scientists predict that China will face some of the worst impacts of climate change' (Marks, ibid.). For Brahm (ibid.), China can hit two birds with one stone by harnessing the sunshine received by the Tibetan Plateau – the strongest sunlight of any place on our planet other than the Sahara – to satisfy its energy needs and deal with a security problem by peaceful means. China is already the world's biggest maker of solar panels (as well as wind turbines) and has plans to turn northwestern Qinghai province into a base for solar power (ibid.).

The overall assessment that can be made in relation to this indicator suggests that the PRC has adopted a positive posture toward ratifying international environmental treaties, albeit subject to those not acting as an impediment to reaching her economic and developmental targets. This can be understood as a continuation of the country's Dengist modernization plan rather than the result of Olympic Games hosting development, which is itself another cog in the developmental sequence of that plan. All in all, the PRC has also scored positively in this EM indicator.

(iv) Protected natural site designation

In an article on environmental law and policy in the PRC, Beyer (2006a, p. 199) argued that

> the leading causes of China's biodiversity loss are extensive agriculture, industrialization, illegal logging and land degradation. About one-third of China's farmland has been exploited from primary forestland and the use of pesticides as well as chemical fertilizers has increased significantly. Despite reforestation efforts, China's overall amount of forest cover has decreased continually and grasslands ecosystems face serious decline. Approximately one-third of the country's deserts are the result of human activity and the trend of desertification is accelerating, especially in China's arid and semi-arid north and north-west. Deserts are increasing at a rate of 1.8 million hectares a year and sandstorms caused by land degradation have become a serious problem in China.

Not surprisingly, the legal framework to address this issue was enacted in 2001 – the year that Beijing submitted its Olympics bid – with the Desertification Prevention and Control Law (ibid.). A related issue is species decline and habitat loss, 'although China ranks second for percentage of nature reserves in the world, no national law on the conservation of nature reserves has been promulgated until now' (ibid.). In addition, one of the environmental goals in Beijing's bid commitments was the 'strengthening of natural preservation zones and the establishment and management of key conservation areas (such as wetlands, forests and bird habitats 8% of the municipal area to be natural protection areas' (UNEP, 2009, p. 15).

Marks (2010, p. 978) offered the following highly informative story line in relation to the EM indicator under examination in this section:

> Every five years, the NDRC drafts and submits to the People's congress a five-year plan (FYP) which sets policy guidelines and targets for each five-year period. Upon assuming responsibility for climate change, the NDRC set guidelines for the first time in its 10th Five-Year Plan (2001–2005) while still placing a greater emphasis on economic development. The plan set targets for vehicles' fuel consumption and energy conservation in residential buildings and public works. These targets, however, were not met (see Richerzhagen and Scholz, 2008). The government did, however, meet its less ambitious target of increasing forested areas, which are carbon sinks, to 18.2%. In the 11th Five-Year Plan (2006–2010), under the broader banner of 'sustainable development', the NDRC set ambitious environmental targets of 20% reduction in energy per unit of GDP and an increase in forested land to 20%.

Similar commitments have been made in the twelfth five-year plan, with an added emphasis 'on protecting arable land, food security, wildlife protection and the "ecological restoration" of areas damaged by construction of roads, rail and other infrastructure. Together, these would mark a turning point for China's environmental degradation' (Watts, ibid.). However, with the failures of the previous five-year plan still fresh, how can we guarantee that the twelfth FYP would not have a similar fate? Indeed, 'while the FYP formation process is becoming more efficient, effective implementation of FYP objectives remain difficult. Local government officials have been known to either slavishly follow plan targets or not follow them at all: during the 11th FYP period (2006–2010) the country's target annual GDP growth was routinely exceeded, while energy intensity targets led to forced electricity brown-outs in several cities in late 2010 to meet those targets. The 11th FYP was also slow in applying fundamental structural changes to the Chinese economy that top leaders say are needed' (APCO, 2010, p. 2). This time, in order to avoid overzealous local officials pursuing a 'growth at all costs' mentality, there is consideration 'to de-link GDP figures from cadres' performance evaluation' (Watts, ibid.). With the positive prospects of the twelfth FYP, this EM indicator also scores positively.

(v) Environmental Impact Assessments

EIAs were officially introduced in China in 1979 through the Provisional Environmental Protection Law as well as the specific protection and conservation laws (Beyer, 2006a, p. 201; Martens, 2007, p. 71). 'However as these provisions set out only general requirements on this matter and particularly lacked implementation measures, a number of administrative regulations and guidelines have been issued over time. In order to unify these highly scattered and overlapping regulations, a new law on environmental impact assessments came into force in 2003 that basically reaffirms and broadens the pre-existing provisions' (Beyer, ibid.).

For Martens (ibid.), the new law 'promotes public involvement in decision-making on projects with potentially adverse environmental impacts. Article 5 [states] that "the state encourages relevant units, experts and the public to participate in environmental impact assessments in appropriate ways". It is further stated in article 11 and 21 that plans and construction projects – subject to an environmental impact assessment – cannot be submitted for approval before the opinions of the public have been solicited through evidentiary meetings, testimony hearings or in other ways'. Nevertheless, Martens (ibid., p. 72) has continued his work by expressing some scepticism about the actual implementation and efficacy of that opening to public participation as follows:

> Although the EIA Law can be considered a step forward in the institutionalization of public participation, several problems with

respect to the inclusion of citizens remain. It is not stipulated how and when the involvement of citizens in evidentiary hearings or testimony meetings should be promoted, leaving this to the discretion of local governments. It remains difficult for citizens to attend public hearings, and even harder to join actively in the discussion. The director of the Chongqing University EIA centre explains: 'During the review of the EIA report we always set two seats for the public, but these people usually remain silent, as the EIA report is very technical and not everybody can understand it.

We can see here an aspect of the EIA process that is not confined to the Chinese case but characterizes its cross-national application as well as its claims of public consultation. It is actually one of the limitations manifested in the cases of the two Olympic hosts examined earlier in this chapter. Having said that, it can still be maintained that '[t]he increased publication of data on local environmental conditions, the elaboration of environmental complaint system and the ongoing learning process with respect to EIA procedures also point in the direction of improved channels for public participation in local environment' (ibid, p. 78). Indeed, emphasizing the limitations of the EIA procedure to engage the average citizen, we cannot dispute the possibilities afforded to citizen groups and ENGOs. For that reason, China also scores positively on this EM indicator.

(vi) ENGO participation

The Environmental Impact Assessment Law (2003) requires public consultation for the initiation of environmental policies and projects. 'Yet, lacking specific prescription is only partially implemented. Constraints also exist on public access to information. Chinese authorities try to control information flows on major environmental issues, especially those of significant political sensibility. Environmental information and news releases to the public lack transparency and interdependency, although the situation is improving' (Xie, 2009, p. 4). For Xie (ibid.), although environmental activism in China generally develops through personal networks, this unravels differently from one region to another. This has been illustrated in her study of environmental activism in four different geographical locations with distinctive characteristics, ranging from the 'very repressive' political system of cosmopolitan Shanghai and the 'rather open' political system found in Beijing, with its 'mixed' local culture, to the 'repressive' political system encountered in the 'traditional' and 'mixed' local cultures of Panjin and Xiangfan (ibid., p. 64). Xie demonstrates how these regime differentials play a key role in the way that personal networks affect the development of the environmental movement in China. Crucially, the Beijing Games was one of the aforementioned environmental issues of high political sensibility and as such, they offer good ground for studying ENGO participation in environmental policy-making processes in China.

One of the commitments made by BOCOG in its bid was to cooperate with ENGOs (UNEP, 2009, p. 16). For Zhang (2008, p. 4), 'BOCOG's dialogue and consulting with NGOs during the Games represent a positive step towards the increased collaboration between government and civil society in China'. Nevertheless, although a number of representatives from ENGOs (Friends of the Earth, Global Village of Beijing) were invited to act as environmental advisors and others (WWF, CI, Greenpeace) were consulted on environmental issues, Greenpeace still emphasised, in its evaluation of the Beijing's green record, that 'channels of communication need to be more frequent and systematic for public participation' (ibid., p. 8).

> There were major challenges to Greenpeace's efforts to assess the environmental achievements of the Beijing Games. Greenpeace's ability to conduct its own comprehensive independent evaluation was constrained by limited access to Olympic venues and comprehensive data [...] Moreover, despite many requests, Greenpeace could mostly only obtain access to information that was already publicly available from BOCOG and relevant government departments. Therefore, some crucial data needed to comprehensively assess the Olympics were unavailable.
>
> (Ibid., p. 6)

Rather surprisingly, considering its status, UNEP (2009, p. 13) reached similar conclusions in its review of the green dimension of the Games:

> In early December 2009, UNEP sent a delegation of experts to Beijing to collect data and hold final discussions with relevant

staff of the Beijing Municipal Government, BOCOG and NGOs. The only comprehensive data that was available on the Games was from the Beijing Municipal Government and BOCOG. This report is therefore based almost exclusively on data from these two organisations. UNEP's team of experts have however, tried to provide objective analysis, comments and recommendations based on this data. It should be noted that in some instances data was not available.

Nevertheless, according to UNEP (2009, p. 121), BOCOG collaborated with more than 30 national and international non-governmental organizations (NGOs) in the organization of the Games over the following issues:

1. developing guidelines for regulating air conditioning;
2. implementing environmental education programmes and raising environmental awareness;
3. acting as intermediaries in the promotion of the environmental activities of BOCOG and government authorities to the media;
4. working with NGOs on specific issues: on the use of timber in the construction of venues (Worldwide Fund for Nature or WWF); as environmental advisors (local NGOs); calculating the carbon footprint of the Games (Conservation International).

However, several ENGOs were concerned that their advice was not taken seriously. In particular, ENGOs 'highlighted their disappointment that environmental measures taken for the Games did not have a holistic approach. Several measures were short-term with no clear indication of what would happen after the Games' (ibid., p. 122). For UNEP (ibid., p. 123),

> [i]t would be useful for the Beijing Municipal government to engage NGOs in discussions on how to carry forward and reinforce the full range of environmental measures undertaken for the Games. NGOs possess specialised knowledge and have emerged from the Games with good experience. The creation of a sustainability forum with representatives from NGOs and various sectors of the public to advise Beijing's Municipal Government could be a positive step.

That is a realistic possibility in the 'mixed' local culture of Beijing, but what about the rest of the country? What would the situation be in Shanghai with a 'very repressive' political system? If we are to be guided by the World Expo mega-event of 2010, ENGO engagement and involvement would be very limited (see Mol and Zhang, 2012). For Xie (ibid., pp. 3–4),

> Besides restrictive regulations on the establishment and development of civil organisations, the state plays a monopoly role in Chinese ecological governance, which provides limited political channels for public participation. At present, relatively few institutional arrangements exist for ENGOs to participate in policy-making and participation on the environment.

In particular, 'only government organised NGOs (GONGOs) and those having established alliances with government people who share their views have some political access' (Lo, 2010, p.1014). This is not necessarily negative. In essence, it is similar to the neo-corporatist policy partnership arrangements encouraged by a few European democracies, which have signified the institutionalization of the environmental movement. After all, Chinese ENGOs had never adopted the radical posture that was characterizing the environmental movement during the 1970s in Western Europe and the United States.

Where were ENGOs placed on the twelfth FYP? Ma Jun, of the Institute of Public and Environmental Affairs, said the government would take a big step forward if it set absolute limits on pollutants and resource consumption, rather than the incremental economy-linked targets seen until now. 'This is the first major effort to translate words into action. Before the government set targets and talked of improvement, but this is the first really major effort to integrate that into an economic plan. They should get credit for that' (Watts, 2011a, p. 29). Furthermore, Li Bo of Friends of Nature said, 'I'm hopeful about the next five-year plan' [...] The government is prepared to go further than before. But we should do more. Until now, low carbon concepts have been introduced only for industry. In the future, I hope those ideas can be adopted in the community so we see a change in lifestyles' (ibid.).

The discussion in this section over participation by ENGOs in the policy-making process has demonstrated the different standing of

Table 7.2   EM in the post-event phase – China

	Beijing 2008
Annual level of CO_2 emissions	☺
Level of environmental consciousness	☺
Ratification of international agreements	☺
Designation of Sites for Protection	☺
Implementation of EIA procedures	☺
ENGO participation in decision-making processes	☺

ENGOs in the Chinese context and their progressive institution-alization in the policy-making process. This process started with the preparation for hosting the Games and has continued to evolve with China adopting FYPs that aspire to delink economic growth from environmental deterioration. The overall impression suggests that the PRC also scores positively in this indicator.

Discussion

China has performed remarkably well in all six EM indicators (see Table 7.2). My assessment could have easily veered in a different direction had it not been for the latest FYP. This positive outcome could be attributed to incremental developments that were bound to take place in China after the 1978 modernizing reforms initiated by Deng Xiaoping. Hosting 'Green Olympics' was an affirmation of this path.

Plans for hosting the Games were implemented with remarkable precision and met or exceeded expectations. With the legacy of the 2008 Games in mind, China followed the requirements of OGI without being required to do so.

8
London 2012: Evaluating the Prospects

The UK is an advanced country that belongs to the G8, and we may expect that it already has a positive score in most, if not all, of the six ecological modernization (EM) indicators. London's should have no difficulty in at least matching Sydney's green standards. However, the impact of spending cuts to key services, made by the coalition government to minimize the country's budget deficit, may significantly affect the current EM standing of the UK and by extension the enriching and amplification of EM capacity that is envisaged by hosting the Games. In its account of these issues, this chapter follows the same strategy that was employed in the preceding chapter to analyse the environmental credentials of three former Olympic hosts.

Starting with an account of the host city, it is important to situate London in the context of global financial flows. For instance, for anthropologist Saskia Sassen, London is a global city, 'a space of power that contains the capabilities needed for the global operation of firms and markets' (2007, pp. 23–4). Fussey and colleagues (2011, p. 77) have argued that 'global cities are considered aspirational benchmarks for the international community and, via their pronounced transformation, set the trends in terms of the organization and internationalization of economic activity, social structure and spatial organization'. Given its nature as a 'space of power' and an 'aspirational benchmark', London is an ideal locus for the examination of the environmental legacy and changed EM capacity that derive from hosting the Olympic Games. This chapter evaluates the prospect that London 2012 will make an effective contribution to the EM of the UK as a whole.

In August 2011, extensive looting and lawlessness, which began in the five Olympic boroughs of Hackney, Waltham Forest, Greenwich, Newham and Tower Hamlets after a protest about the police killing of Mark Duggan, spread across the whole of the UK. This activity adds another parameter to the discussion of the legacy of the London Games. The following paragraphs demonstrate how the aspirations inscribed in the bid for the Games as well as the pre-event preparations have only partially recognized the underlying causes behind the 2011 riots. In particular, what is the impact of the riots on the sustainability claims of London 2012?

For Thornburgh (2011, p. 18), 'it seems to comfort the authorities to see the rioters as amoral outliers [...] But somewhere behind the seeming anarchy lie the very real discontents of modern Britain'. He argues that this discontent is the result of income inequality:

> Start with something called the Gini coefficient, a figure used by economists to indicate how equally (or unequally) income is distributed across a population. In this traditional measure, Britain fares worse than almost every other developed country in the world. According to International Monetary Fund economists Michael Kumhof and Romaine Rancière, nearly 30% of income in the UK in 2005 (the most recent year for which data exist) went to the top 5% of earners; in no other major European country is so much concentrated in the hands of so few. The Organisation for Economic Co-operation and Development says the UK has the worst social mobility of the developed nations: those born to a certain class tend to stay there.
>
> (Ibid., p. 18)

Indeed, the perception of the rioters as 'amoral outliers' was shared by Prime Minister David Cameron, who famously said that the 'riots were not about poverty' (Taylor et al., 2011, p. 4. However, a study conducted by *The Guardian*, which had gained access to detailed datasets about arrests made during the riots, demonstrated 'a correlation between economic hardship and those accused of taking part in [...] violence and looting' (Taylor et al., ibid.).

The riots spread to other cities beyond the Olympic conurbation, yet in terms of geographical distribution, available evidence suggests that in London, 'rioters often looted shops and businesses in or near the areas where they lived' (ibid.). Moreover, an analysis of

> the Guardian's preliminary data by overlaying the addresses of defendants with the poverty indicators mapped by England's Indices of Multiple Deprivation, in small geographical areas, [...] found that the majority of people who have appeared in court live in poor neighbourhoods, with 41% of suspects living in one of the top 10% most deprived places in the country. The data also shows that 66% of neighbourhoods where the accused live got poorer between 2007 and 2010.
>
> (Ibid.)

Other research carried out by the Institute of Public Policy Research (IPPR) 'found that in almost all of the worst affected areas, youth unemployment and child poverty were significantly higher than the national average while educational attainment was significantly lower' (ibid.).

The social demographics of the Olympic boroughs were known long before London was awarded the Games. In fact, those demographics played a key part in formulating the impetus to bid for the Games. A discussion of the actual bid as well as the post-award 'pre-event' phase will help us to understand the crucial role played by this parameter.

London bids for the Games

The awarding of the Games to London was unexpected in both London and the rest of the world. For Evans (2007, p. 298), this decision by the International Olympic Committee (IOC) 'represented a combination of Eurovision-style partisanship, tactical voting, global schmoozing and last minute surprises'. As such, 'land acquisition and relocation had, not surprisingly, been taken less than seriously [...] The regeneration legacy was not reliant upon the Olympics; this would be the icing on the cake and provide the international cachet, even to an established world city and cultural capital' (ibid., p. 307). Indeed, there is little doubt that the rationale for the preparation of the bid involved the hope that the bid would 'encourage further investment in the Thames Gateway from the private sector and if successful also provide a milestone that infrastructure improvements will have to meet' (Vigor et al., 2004, p. 10).

Then mayor of London Ken Livingstone said, 'The Olympics will bring the biggest single transformation of the city since the Victorian age. It will regenerate East London and bring in jobs and massive improvements in transport infrastructure' (ibid., p. 10). Indeed, these aspects constituted a significant element of the actual bid:

> The Olympic Games and Paralympic Games in 2012 will enhance sport in London and the United Kingdom forever. Our people, especially our youth, will benefit from much needed facilities. Our next generation of athletes will be better equipped to develop into future Olympians, reinforcing and strengthening the Olympic movement in the country. Wide-ranging sport programmes will encourage greater participation. The nation will be healthier, happier, and more active.
>
> Throughout our country, there is an appreciation that the Olympic Games and Paralympic Games are a power for good. For London 2012, that power for good will be the most powerful catalyst imaginable for the regeneration of one of our most underdeveloped areas. It will accelerate the most extensive transformation in London for more than a century. Tens of thousands of lives will be improved by new jobs and sustainable new housing, new sport venues and other facilities; all set in one of the biggest city centre parks created in Europe for 200 years. There will be a real and long lasting legacy.
>
> (London 2012, 2004, p. 1)

The UK public overwhelmingly supported the bid, as demonstrated by an Independent Communications and Marketing (ICM) nationwide poll that was conducted in December 2002 that found that '81% said that they thought London should go ahead with the bid itself while 82% were in support of the bid. Nationwide backing was slightly higher. Across the country, the participants from Northern Ireland showed the highest backing with 87% in support followed by Scotland with 84%. There were no geographic areas for which results indicated any type of majority disapproval for the project. In all of the regions polled, the support did not dip below 75%' (Dave, 2005, p. 35).

Poynter (2009, p. 185) claims that London's 'success was attributed to its focus on urban regeneration and the importance attached to the sporting legacy to be provided for generations of young people'. Similarly, in Parliament, Jack Straw, foreign secretary, said,

> London's bid was built on a special Olympic vision. That vision of an Olympic games that would not only be a celebration of sport but a force for regeneration. The games will transform one of the poorest and most deprived areas of London. They will create thousands of jobs and homes. They will offer new opportunities for business and the immediate area and throughout London.
>
> (Ibid., p. 185)

Again stressing the potential for development, Nick Smales, the Service Head for the 2012 Olympic and Paralympic Games, said that '[t]he Olympic Park will be created in the Lower Lea Valley; thirteen kilometres from the centre of London. This area is ripe for redevelopment. The most enduring legacy of the Olympics will be the regeneration of the entire community for the direct benefit of everyone who lives there' (Smales, 2008). The specific demographics that Smales provided made an even more compelling case for the bid in relation to the needs of the population of the five Olympic boroughs (2004).

> By making use of the available Government Indices of Multiple Deprivation (IMDs), we can discern an overall picture of significant disadvantage throughout the Olympic boroughs. A key indicator of this disadvantage is the boroughs' high level of unemployment – 5.12 per cent was the average rate of unemployment benefits claimants for the five boroughs, as compared to 3.3 per cent for London and 2.5 per cent for Great Britain in total (ibid., pp. 25–8). Moreover, the area is marked by 'disturbingly high levels of unemployed residents who have never worked', as well as 'above average levels of long-term unemployed' and educational underachievement.
>
> (Ibid.)

Moreover, according to the IMDs published by the Department for Communities and Local Government (which have not been calculated since 2007), two of the Olympic boroughs, Hackney and

Tower Hamlets, are among the most deprived local authorities in England (Smales, ibid. and *The Guardian* 2011). These demographics are important to bear in mind as we begin to discuss the extent to which sustainable development (SD) commitments made during the bid were actually implemented in 'pre-event' preparations.

The environment in the bid

The London bid (London 2012, 2004) presents a sustainability perspective that manages to combine social and environmental constraints, and it should be noted that such combination has been the traditional aspiration of SD. The bid presented the ambition to regenerate the Olympic boroughs of East London as *the* driving justification for the bid. The bid puts it this way:

> In line with the IOC's Agenda 21, London 2012 is developing a comprehensive environmental and sustainability strategy to optimise environmental protection and enhancement opportunities. Key elements of the strategy will include measuring and monitoring performance across a range of environmental and socioeconomic indicators.
>
> London 2012 will address climate change issues by optimising the most carbon-efficient choices in the Games: use of public transport, rail and river freight; specifying non-polluting official car fleets, buses and service vehicles; energy efficiency in facility design, construction and operation. It will also seek to generate and use renewable energy, and create more green space, wetlands and wildlife habitat. Such actions will also run alongside an Olympic environmental and sustainability awareness campaign and a long-term sustainable sport programme.
>
> (Ibid., p. 23)

A report by the IOC Evaluation Commission (2005) made reference to the bid's 'Towards a One Planet Olympics' concept and to its wider city regeneration plan. Concerns over increasing levels of ozone pollution and the legislative measures and actions that had been taken to ameliorate this situation – the creation of a 'low emission zone' and a 'congestion charge' – also occupied a prominent place in the report, as did the ratification of the Kyoto Protocol.

The 'pre-event' phase: planning sustainable games

Sustainability has also been prominently featured in the post-award 'pre-event' phase of the Games.
 The final, submitted bid proclaimed a commitment to

> [u]se venues already existing in the UK where possible, only make permanent structures that will have a long-term use after the Games, and build temporary structures for everything else.
>
> Aside from using the Games as an opportunity for the regeneration and improvements in the quality of life in East London, the aspiration is to 'encourage more sustainable living across the UK'.
>
> (London 2012, 2007)

London's ambition toward sustainability is not restricted to local and national concerns but moves a step beyond to the global arena. It subscribes to the notion of a global commitment to sustainability. This moves hosting the Games beyond the successful execution of a one-off event into something with a long-term impact.

The London plan has five key themes: climate change; waste; biodiversity; inclusion; and healthy living (London 2012, 2007). Various progress report cards produced by London 2012 appraised these themes systematically. An update and progress report card published in December 2008 (London 2012, 2008a; 2008b) reported satisfactory progress on most of the commitments made in the 2012 Sustainability Plan. Moving forward, the second edition of London 2012's Sustainability Plan (London 2012, 2008b, p. 16) outlined refinements made to its overall carbon management strategy. Importantly, the second edition replaced the original mantra of 'reduce, replace and offset' with four steps 'avoid/eliminate, reduce, substitute/replace, compensate'. This change was more a recognition that a 'carbon neutral' Games was an impossibility than it was a diversion from a full commitment 'to deliver a truly sustainable Games'. Indeed, according to Hayes and Horne (2011, pp. 754–5),

> London 2012 has not set a 'carbon-neutral' goal, has abandoned the highly contentious practice of offsetting, and has developed a carbon footprint methodology calculating emissions 'when they

happen', producing a reference footprint from the point of the bid win to the closing Games ceremony, assuming development as set out in the bid dossier.

Interestingly enough, a final, pre-Olympic Games Impact (OGI) Study report published in October 2010 found 'below average performance for the environmental outcomes indicators'. It continued by suggesting that these indicators 'may be expected to improve as the various environmentally-oriented activities begin to yield results' (ESRC, 2010, p. 25). I expand on the environmental outcomes indicators of this study below. In the meantime, however, since sustainability is not solely composed by the environmental factor, we must devote our attention to the other parameters that the bid itself heavily promoted.

Hayes and Horne (ibid., p. 760) question, 'to what extent a six-year scheme of construction for a four-week festival of sport can rightly lay claim to being "the most sustainable games ever"'. In particular, for concerned parties the sustainability of the Games can only be justified by stimulating social and cultural transformation in the form 'asocially inclusive environmental citizenship' (ibid., p. 760). These questions have acquired added weight in light of the aforementioned riots and their relation to the social, economic and infrastructural problems that the five Olympic boroughs have been facing.

The Olympic boroughs 'have a multi-area agreement and a strategic regeneration framework which puts the emphasis on:

- Creating a high quality city within a world city region.
- Improving educational attainment, skills and raising aspirations.
- Reducing worklessness, benefit dependency and child poverty.
- Homes for all.
- Enhancing health and wellbeing.
- Reduce serious crime and anti-social behaviour.
- Maximising the sport legacy and increasing participation.

(ESRC, 2010, p. 18)

As indicated in Chapter 6, the implementation of the OGI study was a very important development that potentially adds credence to the sustainability commitment of the London Games. This study was 'designed to evaluate the Games' legacy for the host nation and city against a raft of social, economic, cultural and environmental indicators, hence providing an "evidence base" for measuring the positive societal consequences of the Games for its hosts' (MacRury and Poynter, 2009, p. 304). The OGI recommendations were adopted by the IOC in 2001; London 2012 is the second time (after Beijing 2008) that the recommendations have been put into effect. That way, with London 2012 the green standing of the summer Games and the projections that can be made about developments in the ecological modernization (EM) capacity of the host have officially entered a new stage. Which are the specific ingredients of the OGI study behind the London Games?

OGI organized its sustainability analysis of the Games around a triptych of environmental, socio-cultural, and economic parameters. Each of these parameters, in turn, included a set of indicators – 55 in total. The environmental parameter was composed of 11 indicators that included 'both outcomes indicators (looking at changes in the state of the environment) along with indicators measuring certain environmentally-oriented activities'; the socio-cultural parameter was composed of 24 indicators that included 'a mix of social outcome indicators (measuring the state of society) with two different kinds of indicators focussing on sports: one bundle is sports outcomes indicators, and the other is focussed on the 2012 Games themselves'; the economic parameter was composed of indicators that 'cover three different types: economic outcome indicators (measuring the state of the economy), specific outcome indicators for the tourism industry and indicators looking at the finances of the 2012 Games' (ibid., p. 23). Each indicator was ranked according to three characteristics: relevance, rating, and confidence. The indicator was then given a sustainability score by calculating the product of these characteristics.

After identifying these indicators and devising a suitable ranking method, the OGI team had to devise methods to compile this data and determine 'how to assess performance overall in terms of sustainability. In particular [the study] raises the question of how the balance of achievement on these three [parameters] is to be judged' (ibid., p. 19). The team identified two different perspectives for doing so:

1. Performance on all three fronts is necessary for an activity to be contributing to sustainable development.

2. 'A degree of substitution should be allowed for, so that achievement
   in terms of environmental benefits, say, could compensate for lack
   of achievement in terms of economic outcomes (or vice versa)'.

(Ibid., p. 19)

As I mentioned earlier, the OGI report did not produce satisfactory results about the environmental performance of the Games.
Satisfactory results would have meant that there was assurance in the
'available environmental confidence data', but instead, there were
problems with the data that continued 'to depress results' (ibid., p. 25).
These problems are a clear case of a statistical artefact. As such, the
environment will perhaps be better assessed in the 2015 follow-up
study, which will occur three years after the Games. Similarly, the
study of post-Olympics EM capacity that is conducted here should be
viewed in the same context of limitations in the availability of data.

Can we say the same about the socio-cultural and economic factors
of sustainable development? We can think about this question by
employing the two perspectives described above. There were substantial differences in overall sustainability score between the two perspectives. By employing the first perspective, which 'implicitly assumes
that trade-offs are permissible between different dimensions', the
average score was 0.37, whereas employing the second perspective,
which calculates scores for all three dimensions of SD, resulted in the
even lower score of 0.04. Either way, 'with a theoretical maximum
score of 1 and with zero representing the status quo' (ibid., p. 22), for
both cases the scores were low. The authors of the study write:

> In both the environmental and economic cases, [...] figures reflect
> the relatively few areas where it is possible to say with confidence
> that there has been an impact and further an impact that it is
> due to the Games. In the case of the environmental indicators,
> only four out of eleven were considered to have recorded a signifi-
> cant impact during the prime period under consideration and with
> the economic indicators, only nine out of twenty-one. It should
> also be noted that, in relation to the economic indicators, there is
> less certainty about the causality of the Games effect vis-a-vis the
> legacy promises, with 14 of the 21 indicators showing Medium
> rather than High relevance.

(Ibid., p. 22–3)

It is clear from this account that the higher score achieved within the socio-cultural dimension is closely tied to the low scores awarded to the other two dimensions. As such, the critical perspective put forward by Hayes and Horne remains unchallenged by the OGI study.

The study contained in this book is rife with examples that offer additional credence to Hayes and Horne's sense of a 'systemic contradiction of advanced late modern capitalist democracies' that is encapsulated in the 'concept of "sustainable Games"' (ibid., p. 761). The first two chapters highlighted a lack of democratic accountability in the 'imagineering' of the Games. Chapter 7 demonstrated that, even after the first 'green' Olympics (Sydney 2000), there was still evidence that this lack of democratic accountability had continued, albeit in a hidden fashion that involved collaboration with a few select environmental non-governmental organizations (ENGOs). An even clearer case is China. China implemented an OGI study without being obliged to do so, but this did not make China into a democratic polity. Some commentators saw Beijing's hosting of the Games as potentially repeating what had occurred in Seoul in 1988. In that case, hosting accelerated the democratization process of South Korea (see Close et al., 2007; Pound, 2008). For other commentators, however (see Mol 2010, p. 512; Mol and Zhang, 2012, p. 138), the autocratic character of the Chinese regime, coupled with its extensive human rights violations (e.g. Tibet, Falun Gong, the Uighur minority and lack of media openness) demanded the adoption of a more restrictive 'environmental definition of sustainability'. In light of Hayes and Horne's worry that London 2012 has followed only a 'hollowed-out form of sustainable development' (p. 751), is it the case that a similarly restrictive definition of sustainability need be applied?

To understand whether this is the case, we need to look at the specific problems that led Hayes and Horne to draw this negative conclusion. Crucial to the two authors' perspective are the instances of civil contestation that emerged over Olympics-related land use. The Olympic Games have always attracted some form of opposition, and London 2012 was not immune from this. The following section looks into how relevant government institutions dealt with this contestation.

Civil contestation

As we saw in Chapter 2, protest action has been a reoccurring feature of the Olympic Games since their modern reincarnation in 1896. A great part of this contestation takes place during the bidding and 'pre-event' phase of hosting. As far as London 2012 is concerned, 'resistance to the Games has taken a number of forms and, perhaps because of this, has remained fragmented. Moreover, the inability of these groups to articulate precise aims, unite with other disaffected groups or connect their grievances to more powerful narratives has rendered them with limited visibility and longevity' (Fussey et al., 2011, p. 214). Indeed, it appears that an attempt to develop a movement to stop the London bid ('No to London 2012'), which was similar to those developed within other host cities (e.g. the Anti-Olympic Alliance in Sydney; the Anti-2004 campaign in Athens; the Olympic Resistance Network in Vancouver), became defunct soon after the bid was submitted. 'No to London 2012' described itself as a 'coalition of east London community groups, anti-authoritarian and social justice campaigners opposed to bid to stage the Olympics in London in 2012'. In their own words,

> [t]his may be one of our few chances to stop [the] bid in its tracks – the final decision will be made in July 2005. A corporate sponsored, multi-million pound Olympics in London will be a financial and environmental disaster built on lies told to some of the poorest communities in Britain. It will lead to even more draconian 'anti-terror' and public order legislation and be paid for not by the rich or businesses but by Londoners.
>
> (No to London 2012, n.d.)

Judging by this discourse, the group was composed of activists operating in the alter-globalization movement milieu. In total, six different protest campaigns occurred during the 'pre-award' bidding period and during the 'pre-event' phase (see Table 8.1). Some campaigns were of brief duration, and some were still organizing protest events only five months before the opening ceremony of the Games.

I want to briefly discuss two cases of contestation that are included in both Fussey, and Hayes and Horne; these are the Manor Garden

Table 8.1   Contesting London 2012

Group	Issues
NLLDC	New Lammas Lands Defence Committee: Protest against the construction of the Olympic park in Leyton Marshes.
MGAS	Manor Garden Allotments Society: Plotholders resisting their eviction and inferior post-Olympic accommodation.
London 2012 Killing Local Business Campaign	Short-lived campaign by local businesses.
Clays Lane Housing Cooperative	Group organized by residents of Clays Lane Estate to campaign against the compulsory purchase order and their relocation.
TELCO	The East London Communities Organization: A diverse alliance of active citizens and community leaders campaigning on causes intending to benefit the local (East London) community.
NOGOE	No to Greenwich Olympic Equestrian Events: A local community action group contesting the staging of equestrian and modern pentathlon events in Greenwich Park.

Sources: Fussey et al. (2011, Chapter 8); Hayes and Horne (2011, pp. 756–8).

Allotments Society (MGAS) (which was also the subject of a 2009 BBC documentary, *Building the Olympic Dream: The last stand at Stratford*) and the Clays Lane Housing Cooperative. I will also go on to describe, albeit briefly, the other contestations.

MGAS: The Manor Garden allotments were part of a charitable donation made by Major Arthur Vliers to local, working-class families. They were situated in the heart of the proposed Olympic Park. Soon after London was awarded the Games, 'the 80 plotholders learned that they were to be evicted to make way for a concourse between Games venues' (Hayes and Horne, p. 756) such as 'the proposed Velodrome and Handball arenas and the southern areas of the development containing the Olympic Stadium and Aquatics Centre' (Fussey et al., p. 214). The plotholders were evicted in 2007, after the London Development Authority (LDA) was forced to provide

alternative accommodation in Waltham Forest (despite public opposition) as well as relocation within the Olympic Park two years after
the Games (ibid.).

Clays Lane Housing Cooperative: Hayes and Horne (ibid., p.
756) view the Clays Lane housing estate eviction, which involved the
compulsory relocation of 425 residents along with allotment holders
and a traveller community, as a 'direct' case of displacement, similar
to those described by COHRE (2007; 2008). In the case of Clays Lane,
contestation took the form of

> a fragmented lobby containing some who vehemently opposed
> the move and some who advocated only moving en bloc as a
> 'community' but – crucially – there were also residents who read-
> ily accepted the proposed relocation and even some who wanted
> to move as quickly as possible out of fear that they might 'miss the
> boat' in terms of the superior relocation packages on offer from
> the council. Ultimately, as people gradually left the estate, the
> effectiveness of the Clays Lane campaign eroded.
>
> (Fussey et al., p. 715)

According to Hayes and Horne (ibid.), it might be the case that
many residents 'were rehoused satisfactorily', but the community was
effectively dispersed, thus adding significant strain to local authority
social housing lists.

Moving on to the remaining three protest movements, a more
functional organizational outlook has been encountered with con-
comitant differences in mobilization outcomes. The most importance difference is the durability of these campaigns.

NLLDC: The NLLDC emerged just after the announcement of the
London Olympic bid with a series of protest events that called for
the preservation of 'ancient rights to "Lammas Lands" at Leyton
Marshes, East London'. According to protesters, 'Leyton Marshes
have been under the protection of a "Lammas Land" covenant for
over 1000 years that allowed the area to be used as a public space with
unfettered movement and access'. As such, the construction of the
Olympic Park constituted inappropriate land use because its fences
'partitioned the land in breach of the Lammas Land Agreement'
(Fussey et al., p. 214).

Interestingly, the Manor Garden Allotments were to be relocated to Marsh Lane playing fields. Iles (2007) describes this development in the following way:

> In a typical gesture of divide and rule, the LDA has chosen to transplant one problem onto another – LLDC and the MGA Holders are two of the groups contesting the current Olympic plan. However, the allotment holders do not want to move at all from the site they have occupied now for over 80 years, preferring to be incorporated as part of the so called 'green' Olympic site. So the two groups have a common cause. Angry locals and members of the NLLDC contend that the proposed site is common land and therefore, should be preserved as such for the use and enjoyment of all, not by a few.

In terms of resource mobilization, the Lammas Land issue provides a storyline that is very attractive to protest actors who operate in the global justice milieu. Indeed, the ancient rights that it recalls (under a long-forgotten name) provide the movement with the aura of indigenous tribes who are fighting the encroachment of terrible conquistadors. Unsurprisingly, Games Monitor, a network of people who raise awareness about issues within the London Olympic development processes, continued to call for protest mobilizations at Leyton Marshes in March 2012 (Games Monitor, n.d.). Having said that, these mobilizations attracted only a small number of participants.

NOGOE: No to Greenwich Olympic Equestrian Events is another land-use Olympic protest movement. It is still active as of March 2012. Fussey and colleagues do not include NOGOE in their list of groups, perhaps because it is not located within the Olympic Village area. Nevertheless, for Hayes and Horne (p. 757) 'it is striking how both [NOGOE and MGAS] have consistently underlined the Olympic planning proposal as "inauthentic"' as in NOGOE's view there is no tradition of horseracing in the borough and staging these and other events in the Greenwich site would be damaging to both the local heritage and environment. Adding to these environmental costs, the 'London 2012 equestrian venue at Greenwich Park has been exempted from the hosepipe ban that [came] into force on April 5 after concerns were raised that drought conditions could create dangerously "fast going" at the Olympic eventing track' (Cuckson, 2012).

TELCO Citizens: The East London Community Organization was launched in 1996 and in 2002, and its trustees are 'affiliated to the London-wide London Citizens organization' (Fussey et al., 2011, p. 217). 'Amongst other causes, [TELCO] seeks to oppose the capitalistic ethos associated with the Olympic-related transformation of East-London in the hope of attaining community benefit' (ibid., p. 216). TELCO currently 'counts over 250 communities in membership' (TELCO, 2011). 'In comparison to other groups opposing the Games, [it] can be considered a major player in the [social movement (SM)] stratosphere. TELCO is actively campaigning for the equal distribution of legacy benefits, which is crucial to determining whether Londoners will benefit from the Games' (Fussey et. al, ibid, p. 222). Given this situation, bringing TELCO aboard for the London bid was crucial in promoting popular support for social diversity and community involvement. It also offered additional credibility to the bid's SD ambition:

> TELCO felt that any public support they offered should be conditioned upon clearly specified benefits for the East London populace. Consequently, TELCO developed, through member-driven discourse, a charter for an 'Ethical Olympics'. The end product was a member-produced video and document written by TELCO entitled the 'People's Guarantees' presented to the Olympic bid team in 2004 by a selection of TELCO's school-aged members.
>
> (Fussey et al., p. 220)

That document included six key points that targeted socio-structural deficiencies in East London, such as a severe lack of affordable housing and employment opportunity. Lord Coe, (then chair of the Olympic Bidding Committee), Ken Livingstone, and John Biggs (Deputy Chair of the LDA) were among the signatories of the agreement. Nevertheless, the truth is that, irrespective of its importance, the contract was not binding. The planners signed a 'carefully worded' agreement that 'ensured room for negotiation should London win the bid' (ibid., p. 221).

Indeed, an examination of TELCO's activities after the Games were awarded to London is revealing of how promises of democratic accountability can become dead letters in the context of mega-event preparations, where 'the immutability of deadlines, the stakes of the

Table 8.2   The quest for an Ethical Olympics: TELCO's activities in the post-award phase

---

1. Solidified the Ethical Olympics agreement with the relevant parties: Mayor's Office, LDA, Olympic Delivery Authority (ODA).
2. ODA refused to interact and acknowledge TELCO.
3. TELCO organized collective action to achieve ODA recognition.
4. Picket rallies organized outside ODA meetings.
5. Campaign levelled off in 2008.

Source: Fussey et al. (p. 221).

reputations, the primacy of delivery' (Hayes and Horne, p. 761) gain uncontested priority. Table 8.2 presents TELCO's activities during the post-award phase:

Given these case studies and the aforementioned democratic deficiencies that are evident in the 'pre-event' phase, the concept of a 'sustainable Olympics' may be a contradiction in terms. But can a similar line of argument be used in relation to a more restricted definition of sustainability that is closer to the EM perspective? The next section attempts (in a somewhat speculative way, given that the London Games have yet to occur) to demonstrate the impact that austerity measures and Cameron's 'big society' have had on available EM indicators.

Existing EM capacity in the UK

I want to proceed now to an analytical examination of the existing EM capacity in the UK context. On the one hand, London 2012 produced a highly competent plan for the execution of a sustainable Olympics during the bidding process for the Games. On the other hand, recent developments suggest that the country is experiencing significant difficulty maintaining its environmental credentials, or even, for that matter, meeting its targets as a signatory of international environmental agreements and protocols. For instance, in May 2011, Chris Huhne, then Secretary of State for Energy and Climate Change, announced that the UK may have to reconsider its commitments in early 2014. The underlying rationale was that the Emissions Trading Scheme (ETS) of the EU was not as ambitious as the one planned in the UK and, given this situation, the UK's commitments could become an added strain on UK businesses (Stevenson, 2011). In response to this comment, 15 charities dedicated to the environment, wildlife and sustainable development reminded David Cameron of his ambitious pledge, made just one year earlier, to make his coalition 'the greenest government ever'. In their letter to the Prime Minister, these charities praised the significant and positive steps taken by the government in this vein, including its 'decision to set up a green investment bank, a commitment to a Natural Environment White Paper following UK leadership at the biodiversity convention at Nagoya, and the decision to fund the demonstration of carbon capture and storage'. Since that time, however, the groups noted that there have been a number of setbacks, such as the inclusion of all environmental protection law within the so called 'red tape challenge' (FoE, 2011).

What is the existing EM capacity of the UK in relation to selected EM indicators? Any answer to this question must take into consideration the potential impact of the current economic crisis, the accompanying austerity cuts, and Cameron's concept of the 'Big Society'. Any projections about the post-Olympics EM capacity of the UK have to be subdivided into two parts. The first considers the performance of each EM indicator before the coalition government came into power. The second, which for the sake of convenience I label 'Big Society', puts together the critical points and challenging issues that have been raised in relation to this new frame. Before examining these indicators, it is important to engage with the notion of the 'Big Society' and identify its origins and underlying rationale.

The 'Big Society'

The origins of the 'Big Society' can be traced back to the neo-liberal perspective advocated by Margaret Thatcher. Freedom of consumer's choice, deregulation and an increasing role by civil society in what had formerly been provisions of the state were its key characteristics. With regard to the last characteristic and in order to bring out into the open the continuity between Thatcherite ideology and Cameron's 'Big Society' we should mention that, '[f]rom 1979, Conservative administrations gave significant support to the voluntary sector, both rhetorically, extolling the virtues of non-state social action, and in practical terms, through funding initiatives [...] From 1979–80 to 1986–87, public sector direct grant support for the voluntary sector increased by over 90% in real terms' (Hilton et al., 2010). Indeed, the empowerment of civil society as 'a space or arena between household and state, other than the market, which affords possibilities of concerted action and social self-organization' (Bryant, 1993, p. 399) has been, for Kumar (1994, p. 130), something more akin to the anarchist tradition. Indeed, the Thatcherite perspective has also been called 'anarcho-capitalism' (Marshall, 2008).

> The current 'coalition government has pledged to move away from big government to "the big society", in which civil society is to be revived and given a greater role in tackling social problems' (Hilton et al., 2010). Available data suggests that, contrary to pessimistic predictions about the decline of civic participation, 'new social movements, pressure groups and non-governmental organisations (NGOs) have often seen quite spectacular growth'.
>
> (Ibid.)

This changing aspect of civic engagement constitutes the bulk of the critique of the 'Big Society' (Coote, 2010). The following argument is characteristic of this critique:

> Increasing the volume of voluntary action is seen as a way to cut public spending. But that's as far as the 'Big Society' vision goes to address the economic causes of poverty and inequality. It pays no attention to forces within modern capitalism that lead to accumulation of wealth and power in the hands of the few at the expense of others. Nor does it recognise that the current structure of the UK economy selectively restricts the ability of citizens to participate.
>
> (Ibid., p. 3)

Coote (ibid., pp. 4–5) confronts the logic that a smaller state necessarily leads to a 'bigger' society:

> That depends on how small the state becomes and what it does. We don't want an overbearing state that depletes our capacity to help ourselves. But we do need a strategic state that is democratically

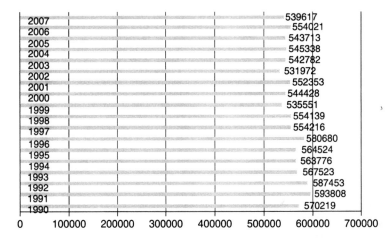

Figure 8.1 UK's CO$_2$ emissions (kt)
Source: UNSD (2010).

'whilst emission cuts achieved in the 1990s will ensure that the Kyoto target will be met, it had long been obvious that the UK would miss its tougher domestic goal of a 20% reduction of CO$_2$ emissions by 2010'. Carter, of course, drew these conclusions under the previous Labour administration. What assessment can be made under the 'Big Society' framework?

In his report to Friends of the Earth (FoE) one year after the formation of the coalition government, Jonathan Porritt (2011) stipulated that, as far as the promise on an 80 per cent cut in greenhouse gas emissions by 2050 was concerned, 'the committee on Climate Change has indicated that Treasury hardliners are opposing the recommended indicative emissions target (of 60 per cent by 2030) based on the difficulties of the current economic situation' (ibid., p. 14). Given this situation, it appears that the UK must be awarded a negative score for this EM indicator.

(ii) Environmental consciousness levels

In a 2008 national survey, 77 per cent of respondents said that they were at least 'fairly concerned' about climate change. At the same time, an equal percentage of respondents 'endorsed the statement that "most people are not prepared to make big sacrifices" to help stop climate change', and a rather large number (60 per cent) questioned the human contribution to climate change and 'thought the government is using the climate change agenda to raise taxes' (Giddens, 2009, p. 101). Despite this scepticism, and even after the online release, in November 2009, of a series of emails between climate researchers that appeared to give credence to suggestions put forward by the climate 'sceptics' community 'that the scientific basis for global warming was flawed', 83 per cent of Britons agreed that climate change is 'a current or imminent threat' (Carrington, 2011). Moreover, 68 per cent agreed that 'humanity is causing climate change', representing only a 3 per cent decrease from August 2009 responses. According to a 2009 global public opinion poll conducted by the University of Maryland's Programme of International Policy Attitudes, 73 per cent of Britons wanted higher priority for climate change policies, making them among the most enthusiastic about action to stop global warming (Goldenberg, 2009).

The 2009 Eurobarometer (72.1) on European Attitudes Towards Climate Change shows that 45 per cent of the UK public considers climate change to be the 'most serious issue currently facing the world as a whole'; 49 per cent consider poverty, lack of food and drinking water to be the most serious issue; and 45 per cent international terrorism. These percentages are significantly lower than the majority that picks 'a major global economic downturn' (55 per cent)

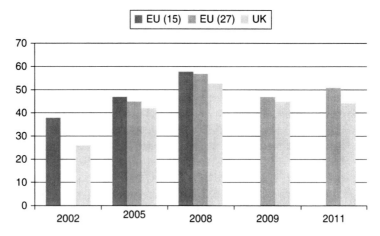

Figure 8.2 Expressed concern about climate change: EU and UK
Source: European Commission (2011).

(European Commission, 2009, p. 10). It is interesting to see how public attitudes have been affected by the economic crisis, which spread in 2010 and 2011. Indeed, the 2011 special Eurobarometer (327) on European Attitudes Towards Climate Change demonstrates an overall increase of concern across the EU's 27 states, with 51 per cent considering climate change as 'the most serious problem facing the world as a whole'. That is a 4 per cent increase since 2009, but below the 57 per cent that was scored in 2008. The UK public exhibited a slight decrease in its 2009 score on the same issue, with 44 per cent of the country concerned about the climate. To compare this to the earlier survey, 51 per cent listed 'poverty, lack of food and drinking water' and 45 per cent picked international terrorism, while the 'economic situation' item scored only 39 per cent. While much can be said about how changes in wording can influence survey results, the fact is that the worst aspects of austerity measures had not yet manifested themselves at the time of this survey, nor was the public yet bombarded with news about negative developments in the Eurozone.

During the 1990s and first decade of the 2000s, the British public exhibited high levels of environmental consciousness (and this concern is accompanied by environmental knowledge, instead of the extremely high and unqualified concern that subsides when confronted by situations that affect material well-being, as has been demonstrated in Greece). This concern reached a height in 2008, when 'all three major political parties had already identified climate change as the biggest threat facing Britain, and all advocated ambitious targets for cutting carbon dioxide emissions' (Rootes, 2009, p. 20). In a further development, the Greens managed to send their first MP in to the House of Commons in 2010, despite an electoral system that is notoriously favourable to the two main political parties.

There is little doubt that the UK public will continue to express relatively high levels of environmental concern, although this concern may exhibit fluctuations that depend upon the prominence of various issues in the national and global arenas. Given this situation, this EM indicator achieves a positive score in both columns in Table 8.3 (see below).

(iii) International treaty ratification

The UK expressed 'commitments to sustainable development in the wake of the 1992 Earth Summit and generated several White Papers

Table 8.3   UK's previous EM capacity and the 'Big Society'

	UK's previous EM capacity	'Big Society'
Annual level of CO_2 emissions	☺	☹
Level of environmental consciousness	☺	☺
Ratification of international agreements	☺	😐
Designation of sites for protection	☺	😐
Implementation of EIA procedures	😐	☹
ENGO participation in decision-making processes	😐	😐

in the 1990s, but little of policy substance' (Dryzek et al, 2003, p. 181). Since then, the UK has made remarkable steps on this front. As Schreurs and Tiberghien (2007, p. 25) write, 'the Germans and the British have quite consistently taken on climate change leadership roles within Europe when they have held the council presidency (2005 for Britain and 2007 for Germany).' Nevertheless, austerity policies have been accompanied by resistance to any environmental measure that is perceived to operate against the economic growth of the country. Given this sitaution, any environmental treaty ratification is bound to be subject to challenges similar to those that confronted the ratification of the Kyoto Protocol. That way, this EM indicator cannot be awarded with absolute certainty a score in the 'Big Society'context.

(iv) Protected natural site designation

As of 2006, there were over 356 National Nature Reserves (NNRs) in the UK (UKELA, 2008). In October 2010, the coalition government 'confirmed plans to sell off as much as 150,000 hectares of forest and woodland in England in the biggest sale of public land for nearly 60 years' (Vidal, 2010). That plan was ditched on 17 February 2011 after fierce public opposition. This reaction illustrates the high regard in which the UK public holds designated natural protection areas.

The 2011 White Paper on the natural environment (Defra, 2011) adopts a position that acknowledges the interdependence with and importance of ecosystems for economic planning. This ambitious plan will fail only if the resources necessary for its materialization are not provided. Interestingly, its discourse fits alongside SD and EM perspectives, while at the same time its vocabulary (about 'freedom of choice' and 'personal' or 'communal level' responsibility) fits alongside that of the 'Big Society'. Still, under the general conditions of austerity cuts and significant strains to public services, some have expressed scepticism about its viability.

The November 2010 announcement of increases in funding for Higher Level Stewardship (HLS), a scheme that pays farmers to deliver significant environmental benefits, was well received by both environmental and education groups. 'However, there are concerns that cuts to Natural England, which provides advice to farmers on stewardship, will affect the scheme's overall viability' (Porritt, 2011, p. 25).

Given this situation, although prior to the onset of austerity cuts and the advance of the 'Big Society' context the UK would have been awarded a positive score on this indicator, there are doubts about its capacity to maintain this standing under the new political framework.

(v) Environmental Impact Assessments

Environmental Impact Assessment (EIA) was first introduced in the UK in 1985 and it was the subject of diverse criticisms that ranged from questions about its usefulness to its ability to conform to the existing planning systems. This initial trepidation has slowly disappeared as more people have been trained in the practice of EIA (Weston, 2002). According to Fischer (2010: 62) '[s]trategic environment assessment (SEA) has been a legal requirement in spatial and other sectoral plan and programme making in European Union member states since 21 July 2004, following implementation of Directive 42/2001'. Fischer (2010) 'performed quality reviews of 117 English spatial plans core strategy SEA reports' and identified a number of shortcomings in its practice (ibid., pp. 62–9). These shortcomings appear to conform to the usual pattern limitations identified in the application of EIAs and SEAs in general and those performed by Olympic host nations, under immense pressure to bypass to meet deadlines for project delivery, in particular.

The white paper on the natural environment that was presented to the House of Commons by Secretary of State for Environment, Food and Rural Affairs Caroline Spelman in June 2011 made a strong case for making the environmental factor a key component of the

'Big Society' by assessing its impact in all planning procedures and arrangements:

> Through reform of the planning system, we will take a strategic approach to planning for nature within and across local areas. This approach will guide development to the best locations, encourage greener design and enable development to enhance natural networks. We will retain the protection and improvement of the natural environment as core objectives of the planning system. We will establish a new voluntary approach to biodiversity offsets and test our approach in pilot areas.
>
> (Defra, 2011, p. 3)

In the same document, the government pledged that it would 'fully consider the value of nature in all relevant Impact Assessments' (ibid., p. 43). These pledges at least appear to subscribe to Cameron's own green ambitions by complementing existing EIA policies. However, since EIAs are very much the responsibility of local authorities and these authorities have been the recent recipients of significant budget cuts, there are doubts about the current efficacy of EIA. In the the FoE's review (Porritt, 2011 p. 48), it is stipulated that the decision to dispense with the audit commission 'has had a big impact on the readiness and ability of Local Authorities to deliver sustainable outcomes at the local level. The Comprehensive Area Assessment has been swept way, together with all the National Indicators that previously helped define the relationship between national and local government. Many of these indicators have a strong sustainability element'. Given this situation, we are forced to mark this EM indicator with an uncertain symbol.

(vi) ENGO participation

According to Dryzek and colleagues (2003, p. 180), during 'the 1990s, environmental groups could re-enter the corridors of power as the excesses of market liberalism eased and the British state reverted towards a more passive orientation, mixing inclusion and exclusion'. However, this 'inclusion often acted as a burden upon environmental groups and [...] doubt[s] on the value of sustainable developments in the United Kingdom' were also raised (ibid., p. 181). Furthermore, although 'belated moves in the direction of ecological modernization

can be discerned in the United Kingdom, beginning in the late 1990s
[...] any movement from weak to strong ecological modernization
requires an active oppositional civil society, capable of engaging
sub-politics in Beck's terms' (ibid., p. 183). Rootes (2009, pp.
17–8) suggests that, despite overwhelming acceptance of the seriousness of
climate change by politicians of all colours, the scope for mobilization by ENGOs around this issue is limited:

> How should ENGOs conduct themselves at a time when environ-
> mental issues have moved to an unprecedentedly central position
> on the political agenda? There are ironies in the present situation.
> Opportunities of access to lobby and potentially to influence
> policy formation have never been greater, but, especially when
> viewed from the perspective of the enormity of the environmental
> challenges confronting us, the yield in terms of policy output
> and effective government action has so far been disappointing.
> Democratically elected governments worry that effective action
> on climate change will incur costs for which the public is unpre-
> pared, and government ministers now urge ENGOs to mobilise the
> public to demand the policies that governments know they must
> implement. Yet governments baulk at providing ENGOs with the
> resources necessary to undertake any such mass mobilization.
>
> (Ibid., p. 27)

If this was the prevailing situation under New Labour, what is the
situation under the current coalition government with its austerity
measures, its ideology of the increased role of civil society actors, and
its purported interest in green government and a green Olympics?

I noted earlier that the 'green economy' concept permeates both SD
and EM. 'In opposition, both the Liberal Democrats and Conservatives
showed signs of understanding the need to connect the two [envi-
ronment and economy], as seen, for example, in George Osborne's
pledge to establish a Green Investment Bank or Vince Cable's plan
to transform aviation taxation' (Porritt, 2011, p. 7). How was the
government's relationship to ENGOs shaped with the advent of the
new coalition? Cameron's 'Big Society' discourse appeared to herald
times of increased participation in decision-making processes, and
this was seen as 'entirely compatible with the type of progressive,
radical emphasis on decentralisation and civic empowerment that

the Green movement has been advocating for many decades' (ibid., p. 37). In the case of the ENGO indicator, however, criticisms have been raised about the capacity of the 'Big Society' to fulfil its aspirations in the context of austerity cuts to public services. For instance, the closing of the UK's oldest ENGO, Environmental Protection UK (EPUK), which was due to cuts to local authorities (Webb, 2011), is perhaps illustrative of what is to come for many other civil society groups and organizations in the UK.

Discussion

Prior to the development of the 'Big Society', the UK scored positively on three EM indicators and it had an ambiguous score on the other three. The only indicator on which it has continued to score well in the 'Big Society' context is 'level of environmental consciousness' (even if this indicator is subject to a range of qualifying factors such as the fact that this concern was accompanied by high levels of environmental knowledge and increased support for green party formations). This study reveals that the other two indicators on which the country scored positively during the pre-'Big Society' period have since been downgraded to the realm of ambiguity, whilst the two indicators that previously had ambiguous status have since been downgraded to negative. The indicator 'annual level of CO_2 emissions' became negative because the highly ambitious reduction targets that had been set by both the current and preceding UK governments are now highly unlikely to be met; the 'ratification of international agreements' indicator was downgraded to ambiguous status in light of scepticism about and polemic against international agreements that are now seen as detrimental to the UK's financial interests. The 'implementation of EIA procedures' indicator was downgraded because of the impact of austerity cuts on the ability of local authorities to carry out these procedures effectively. Austerity cuts were also blamed for the fact that the 'ENGO participation in decision-making processes' indicator remained ambiguous. Most importantly, this took place in the context of what appeared to be an ideal situation for the demands of the green movement. All in all, under the 'Big Society', the UK's EM capacity was downgraded from an overall score of 3/6 to 1/6.

Concluding remarks

In an earlier commentary (Karamichas, 2012b, pp. 390), I argued that 'the UK had been in an advantageous position in catering for the environmental legacy of London 2012 when compared to the position that the two modern Olympiads that followed Sydney 2000, the first 'Green Olympics', found themselves'. As far as variables conducive to EM are concerned, we can see that the UK has traditionally occupied very much a leading position. Whereas studies on the environmental legacy of other host nations, and specifically accounts of the extent to which Olympic hosting can be used as an opportunity for the EM of the host nation, the EM that accompanies London 2012 should be largely a *fait accompli*. However, certain negative dimensions of the political process that have affected the EM process in other host nations may also play a role in hampering the maintenance and development of EM in the UK.

Indeed, earlier prospects suggested that the Olympic Games would complement the UK's existing EM capacity. The 'Big Society' and green rhetoric of the coalition partners raised these expectations. Nevertheless, when this rhetoric was complemented by unstoppable waves of austerity cuts, it became apparent that the country risked losing its previous EM status after hosting the Games.

9
Concluding Remarks

This work brought under the same roof two issue-areas. One of them dealt with the most popular sport mega-event, the Olympic Games, and the other dealt with the environment, in all of its expressions. In essence, this book has been dedicated in its entirety to an examination of the inter-relationship between these two issue-areas by employing the instruments of critical analysis offered by environmental sociology. The initial impetus for this exploration was stimulated by the identification of a developmental process whereby both of these seemingly incongruent items converged with one another to such an extent that mega-event hosting without factoring for the environment had become inconceivable. With the emergence of contemporary environmental concern in the 1960s and 1970s, the staging of the Olympic Games and the protection of the natural environment and its resources from over-exploitation appeared to be guided by diametrically opposed rationales and processes, yet since the 1990s, prospective Olympic hosts have been encouraged to factor in the environmental dimension in both their bidding application and in their subsequent preparations for hosting the event. That development further intensified the initial impetus by opening up a new, exploratory rubric that called for a full-blown exploration of the capacity of Olympic Games hosting in affecting the ecological modernization (EM) of the host nation. The result was a cross-national comparative study that examined the environmental standing of each host city and nation during the different phases in the development of its Olympic edition. The selection of EM as a theoretical tool, which simultaneously possesses a normative and prescriptive capacity, called for an exploration of the disciplinary research that gave birth to the EM perspective. In effect, this has been seen as a win-win situation, whereby both the sociological sub-discipline of environmental sociology in general and the reflexivity perspective of EM in particular have both benefited. After all, in both cases we have a clear manifestation of their immense value as contributors to the theoretical development of the discipline and, by extension, the aptitude of the discipline to deal with real world issues. That is of immense value, especially in the context of austerity cuts under which the humanities and social sciences have been under constant attack.

The role of environmental sociology

The guiding inspiration of work about a dysfunctional marriage between the Games and the environmental issue was further encouraged by a recognition of the need to identify the contribution that the environmental sociological imagination can make to the discussion of these two issue-areas as well as simultaneously demonstrate how environmental sociology can be furthered as a sub-discipline. Just as the earlier perception that sport mega-events were incompatible with environmentalism appeared to be outmoded, so the relationship between the sociological and the environmental dimension followed a similar developmental path; therefore, the emergence and development of environmental sociology from its initial skirmishes with classical sociology to its apex of reflexivity had to be accounted for. The book accomplished this by forming an account that began with the emergence and development of environmental concern. It then bifurcated into a dual passage occupied, on the one hand, by a developmental sequence that culminated in the 'greening' of the Olympics and, on the other, by bringing the environmental dimension to sociology as a core stimulant in its theoretical endeavour as well as the main ingredient for assembling a model of analysis that would facilitate the testing of the environmental credentials of four Olympic host nations, beginning with Sydney 2000 and culminating in London 2012.

This move was stimulated by the well-established perception that Sydney was the first city to bid for the Games on a platform of green promises that subsequent hosts then had to emulate. Although Athens, the host city that followed, did not hesitate to follow closely the green promises made by Sydney in its bidding application, it failed miserably to put them into practice. The extremely negative reports by prominent environmental groups on the failure of Athens 2004 to meet its green promises posed the following question: 'What makes some nations run successful green Olympics and others to fail miserably in this direction?' By posing this question, I entered a process that called for a multivariate and cross-disciplinary analysis of two completely different nation-states. This project was further accentuated when this initial two-country investigation expanded with the addition of two more host cities, Beijing and London. This added to the complexity of the study since, with the addition of Beijing 2008, the research was obliged to engage in the study of a non-Western country; with the addition of London 2012, it had to apply a framework of analysis that was formulated to assess the 'post-event' EM capacity of a city that was then in the 'pre-event' period.

The analysis of the London Games was complemented by two developments: the application of the first official Olympic Games Impact (OGI), which put under the microscope the sustainability credentials of the Games; and significant austerity cuts and their impact upon the EM of the UK. The economic crisis that generated these austerity cuts has brought the analytical debate back to the issues that were identified in relation to the Greek case. Developments in the Greek case inspired the employment of two descriptors that were then used to facilitate the discussion of the EM capacity of Olympic hosts in the context of the global economic crisis. The second descriptor, 'canary in the coalmine', viewed the Greek case as signifying the general spread of the economic crisis to other national contexts. The underlying rationale was that the impact of austerity measures in Greece would be replicated elsewhere in the world. Although Greece has a substantially different structural framework than that of the UK, the examination of the impact of the austerity measures by the coalition government in the study of London 2012 at least partly supported the logic of that descriptor.

IOC and the environment

What was the precise nature of the green perspective advocated by the International Olympic Committee (IOC)? That was another important question that I had to tackle in order to offer an answer to the core question behind the project. In this process, I consulted the environmental section of all of the Manuals for Candidate Cities (MCCs) that guided the bidding applications of the four successful hosts. Leaving the London 2012 MCC aside (because London is the first Summer Games host city to be mandated to carry out an OGI study), I found that, other than small changes of wording, there were not any substantial differences among the four MCCs. Nevertheless, this general similarity did not stop these prospective hosts from responding to the requirements of the MCCs in different ways. Their responses ranged from evidence-based appraisals of existing capacities to over-ambitious claims, which, although admirable in their proclaimed intentions, had little realistic prospect of being met.

This was the predominant characteristic of the Greek case. Indeed, the 'pre-event' phase of Athens 2004 was marred by a substantial number of difficulties that led to increased financial costs, delays in project delivery and significant diversions from the city's original plans. In that case, the environmental issue was bound to be negatively affected by these problems, and this fact is clearly shown in the environmental assessment of Athens 2004 by leading environmental non-governmental organizations (ENGOs). A few positives were, however, identified, and these substantiate some sort of environmental legacy that has been bequeathed by Athens 2004. Most of these are related to transport-related infrastructural changes and the beautification of the Athenian centre, but there was also the successful increase of environmental awareness in the general public.

This allowed us to include the Greek case as part of the overall attempt to provide an answer to the following question: 'What contribution towards the ecological modernization of the host nation can be made by Olympic Games hosting?' The phrasing of the question exemplifies the commonality between the proclaimed aspirations of the IOC to include the environmental factor in awarding the Games and the analytical vigour that EM, as an environmental sociology perspective, can offer in answering this question. This, however, is not a straightforward and uncontested linkage. As it was claimed in the introductory chapter, the sociological discipline was an uneasy bedfellow with both the environment and sport mega-events. The first six chapters undertook the task of unravelling the threads that underpinned the above rationale. The conceptual schema that was employed was constituted by three elements: sociology, the environment (concern/behaviour) and sport mega-events (Olympics). With the environment as a centrifugal separator, Chapter 3 was dedicated to tracing the emergence and development of environmental concern that was paralleled by the development of environmental sociology detailed in Chapter 4. That chapter dealt with an overview of the placement of the environment by classical sociology. It also identified the misconceptions that partially guided the founders of environmental sociology and offered some insights into the Marxian-inspired political economy and eco-socialist perspectives. The latter aspect was not only seen as an element of my quest but as a necessary counterweight for balancing the detailed engagement with and coverage of reflexivity that was performed in Chapter 5. In essence, that chapter provided closure to the aforementioned parallel between the emergence and development of environmental concern and environmental sociology by showing how an unprecedented environmental disaster, the 1986 Chernobyl nuclear accident, led Ulrich Beck to found the environmental dimension in sociology. Beck advocated a reflexive modernization (RM) perspective alongside new perceptions about trust in established authorities, scientific reasoning and protest politics. However, the rather depressing perspective that guided it was soon followed up by the similar, albeit different perspective offered by ecological modernization. EM was similar to RM in advocating the need for the further modernization of modernity, but it was different in that it did not adhere to the highly critical perspectives that were adopted by Beck. Most importantly for the purposes of this study, the fact that EM was in proximity to the aspiration put forward by the IOC in relation to the environmental benefits of Olympic Games hosting led me to close Chapter 5 with an engagement with the scholarly debate on EM-capacity building. This was followed, in Chapter 6, by a brief account of the way in which the IOC had factored the environmental issue into its operations and how it has evolved plans for the long-term legacy of the Games. That paved the way for a full-blown exploration of the extent to which the organization of successful Green Olympics is a simple, one-off event or a lasting platform that can diffuse its standards across a range of economic and technical activities within host nations. This was in turn performed in Chapter 7 by using the six EM indicators that I identified in my examination of the EM perspective in Chapter 5.

The following issues became apparent in the examination of three Olympic hosts in Chapter 7. There are good reasons not to accept the green credentials of Sydney 2000 in their entirety. To be sure, Sydney set a good precedent for following Olympic editions, but any encomiums are likely to miss the fact that Sydney did not completely cross the Rubicon of understanding the expenses of hosting the Games and achieving a sustainable legacy on both the social and environmental fronts. As such, the overarching failure of Athens on the environmental issue did not automatically award immunity to Sydney 2000 in the re-examination of its green credentials. Moreover, it was demonstrated that the engrenage ambitions envisaged by the IOC have been blocked in both cases by the political dimension. In the Australian case, that blockage was initially caused by the narrow-mindedness exhibited by the Howard government in refusing to ratify the Kyoto Protocol and in keeping ENGOs away from the policy-making process; in the Greek case, it was caused by the intransigence that was exhibited over environmental issues by the Karamanlis government. In this way, Andersen's second hypothesis, that 'the degree of ecological modernization in a particular country depends on its capacity for environmental reform as fostered and supported by the character of the political and socio-economic reform process' (Andersen, 2002, p. 1396) has been supported by both the Australian and Greek cases.

All in all, no causality was identified between Olympic Games hosting and improvements in the EM capacity of the host nation. Developments in this direction were tied to political changes. For instance, the 2007 election of the Australian Labour Party (ALP) in Australia, which was closely followed by the immediate ratification of the Kyoto Protocol, and the 2009 election of PASOK (Panhellenic Socialist Movement) in Greece, which was accompanied by a range of highly ambitious plans for the EM of the country, were characteristic examples of the significant role that can be played by changes in political administration. Nevertheless, in both cases, perceived threats to the economy took precedence over the successful meeting of requirements for achieving a positive score in the six EM indicators. In the Australian case, we had the internal demise of the Rudd government and a decrease of support for ALP due to delays in the implementation of a number of popular climate-change-related measures; in the Greek case, we had the replacement of the minister of the environment by the minster of economics (who was also an unpopular orchestrator of many of the severe austerity measures). The optimism that characterized the closing remarks that I made in my earlier treatment of the two cases (Karamichas, 2012a) has been significantly undermined by the advent of the economic crisis that hit Greece, as well as the inability of the ALP government to afford losing support by its blue-collar voters. In this way, Andersen's second hypothesis continues to be supported.

Moreover, the assessment of the Greek case has brought into the fore the first descriptor, 'blessing in disguise'. With all the sadness that one can feel about its impact on the welfare of the Greek population, we cannot disregard the positive developments that may come out of the radical restructuring that Greece has been forced to make in order to meet the demands of its lenders. In a nutshell, these developments are intimately connected to the political and economic modernization of the country and in that way the aforementioned descriptor makes perfect sense. As I have argued elsewhere (Karamichas, 2012a, p. 171), Greece has been 'dominated by characteristics that are not conducive to modernization, let alone its ecological dimension; in order to achieve this, changes in the "software" may have to be so encompassing that they have a serious potential to be seen as changes in the "hardware"'. These all encompassing changes to the 'software' have a greater chance of being materialized under the conditions now enforced by the current crisis.

Certain notable improvements were identified in the Chinese case, but these were more a part of incremental developments that were bound to take place in China after the 1978 modernizing reforms that were initiated by Deng Xiaoping. Hosting the Olympics was simply an affirmation of that path. The latest five-year plan shows a strong ambition for the complete greening of the Chinese economy without compromising the fulfilment of its economic and social goals.

The UK case, which was examined in Chapter 8, initially showed good potential for complementing its existing EM capacity after the Games. Nevertheless, the analysis demonstrated that significant doubts have now been raised over the ability of the country to maintain is positive status in at least half of the identified EM indicators.

The seriousness of this situation, which follows OGI recommendations, will be better assessed in 2015.

This project has been conceived as an ongoing work. It aspires to develop a framework that can be used to examine the EM capacity of nations that host the Olympic Games and other sport mega-events. The next challenge in the pipeline is to apply this study to the Rio 2016 Olympic Games. Brazil, of course, is one of the BRICS (Brazil, Russia, India, China, South Africa) countries.

Notes

2 The Olympic Games: A Quintessentially Modern Project

1. 'Alter-globalization' is a term (along with 'global justice') that has been used to describe the antiglobalization movement.

4 Environmental Concern and Environmental Sociology: Parallel Developments

1. This chapter draws heavily from the first chapter of I. Botetzagias and J. Karamichas (2008), *Environmental Sociology* (Athens: Kritiki). As part of an agreement between the two editors/authors of that volume, each chapter has been written by the corresponding author. As such, and honouring the agreement between the two authors, the corresponding bibliographic reference is simply Botetzagias (2008b).

5 Reflexive Modernization: Connecting the Environment with Modernity and Modernization

1. Indeed, Buttel (ibid., p. 327) mentions that studies have confirmed that the Kuznets curve is correlated with positive environmental impact in relation to the reduction of GHG emissions. By extension, this confirms the view of the post-materialist thesis, namely that income increase in the general population contributes to the satisfaction of basic needs, therefore creating the possibility for the articulation of more qualitative demands such as environmental protection. Buttel also takes into consideration more sceptical accounts such as that offered in the study conducted by Opschoor (1997), which 'noted that the process of delinking economic and income growth from the demands of the biosphere for materials and services in the industrial countries was too slow to yield a Kuznets curve-type response [... and] that the more recent evidence suggests that there may have even been a delinking process between income, pollution, and resource consumption among rich nations in the early 1990s' (ibid.).

7 Olympic Games and Ecological Modernization: Sydney, Athens and Beijing

1. The treatment of the Sydney and Athens Olympiads draws heavily from Karamichas (2012a). However, both cases have been substantially updated

to accommodate the changes that have taken place in Greece with the onset of the global economic crisis. The updates include an appraisal of the magnifying effect of the crisis on existing problem areas in the general socio-political front and, the potential for EM capacity. In the Australian case the updates refer to significant developments that have marked Australian politics after the replacement of Kevin Rudd in the Australian Labor Party (ALP).

Bibliography

Albrow, M. (1990) *Max Weber's Construction of Social Theory* (New York: Macmillan).

Almond, G. and S. Verba (1963) *The Civic Culture* (New Jersey: Princeton University Press).

Altshuler, A. A. and D. E. Luberoff (2003) *Mega-Projects: The Changing Politics of Urban Public Investment* (Washington, DC: The Brookings Institution and Lincoln Institute of Land Policy).

Andersen, M. S. (2002) 'Ecological Modernization or Subversion? The Effect of Europeanization on Eastern Europe', *American Behavioral Scientist*, 45/9, 1394–416.

Andranovich, G., M. J. Burbank, and C. H. Heying (2001) 'Olympic Cities: Lessons Learned from Mega-event Politics, *Journal of Urban Affairs*, 23/2, 113–31.

Androulidakis, I. and I. Karakassis (2006) 'Evaluation of the EIA system Performance in Greece, Using Quality Indicators', *Environmental Impact Assessment Review*, 26, 242–56.

Metrogreece (2011) 'Message by Papaconstantinou before the Durban Conference', http://www.metrogreece.gr (home page), date accessed 28 November 2011.

ANP (2010) *Australian National Parks*, http://www.australiannationalparks.com/, date accessed 10 June 2010.

APCO (2010) *China's 12 Five-Year Plan: How It Actually Works and What's in Store for the Next Five Years*, http://www.apcoworldwide.com, date accessed 15 January 2011.

Apostolopoulou, E. (2009) *The Social Conflicts During the Implementation of Conservation Policy in Protected Areas: Analysis and Appraisal of Conservation Policies in Greece*, Doctoral Dissertation (Thessaloniki: Aristotle University of Thessaloniki, Faculty of Science, School of Biology, Department of Ecology).

Arendt, H. (1958) *The Human Condition* (Chicago: University of Chicago Press).

Asimakopoulos (2009) 'The Olympic Village', presentation from conference *The Impacts from Olympic Games Hosting in Beijing, Athens, Sydney, Barcelona*. http://library.tee.gr/digital/m2482/m2482_contents.htm, date accessed 12 April 2010.

ASSDA (2009) *Australian Election Study, 2007*, http://nesstar.assda.edu.au/webview/index.jsp?object=http://nesstar.assda.edu.au/obj/fCatalog/Catalog17, date accessed 15 January 2010.

ASSDA (2010) *Australian Election Study, 2010*, http://nesstar.assda.edu.au/webview/index.jsp?study=http%3A%2F%2Fnesstar.assda.edu.au%3A80%

2Fobj%2FfStudy%2Fau.edu.anu.assda.ddi.01228&v=2&mode=documentati
on&submode=abstract&top=yes, date accessed 17 March 2011.

Astbury, S. (2011) 'Forest Fires Rage in Australia's West', http://www.newsob-server.com (home page), date accessed 11 March 2011.

ATHOC (1996) *Athens 2004: Candidate City* (Athens: ATHOC).

Australian Bureau of Statistics (ABS) (2006a) *2006 Year Book Australia: A Comprehensive Source of Information about Australia* (Canberra: ABS).

Australian Bureau of Statistics (ABS) (2006b) 'What Do Australians Think About Protecting the Environment?', *2006 Australian State of the Environment Committee* (Canberra: Department of Environment and Heritage).

ASSDA (2010) *ANU Poll 2010: Australia's Future*, http://nesstar.assda.edu.au/webview/index.jsp?study=http%3A%2F%2Fnesstar.assda.edu.au%3A80%2Fobj%2FfStudy%2Fau.edu.anu.assda.ddi.01203&v=2&mode=documentation&submode=abstract&top=yes, date accessed 18 April 2011.

Bulkeley, H. (2001) 'Governing Climate Change: The Politics of Risk Society?', *Transactions of the Institute of British Geographers*, 26/4, 430–47.

Barry, J. (2005) 'Ecological Modernisation' in J. S. Dryzek and D. Schlosberg (eds), *Debating the Earth: The Environmental Politics Reader*, 2nd edn (Oxford: Oxford University Press), pp. 303–21.

Barry, J. (2007) *Environment and Social Theory*, 2nd edn (London and New York: Routledge).

Baudrillard, J. (1995) *The Gulf War Did Not Take Place* (Bloomington and Indianapolis: Indiana University Press).

Beck, U. (1992) *Risk Society: Towards a New Modernity*, M. Ritter (trans.) (London: Sage Publications).

Beck, U. (1994) 'The Reinvention of Politics: Towards a Theory of Reflexive Modernization' in U. Beck, Giddens, A. and S. Lash (eds), *Reflexive Modernization: Politics, Tradition and Aesthetics in the Modern Social Order* (Cambridge: Polity Press), pp. 1–55.

Beck, U. (1995) *Ecological Politics in an Age of Risk*, Amos Weisz (trans.) (Cambridge: Polity Press).

Beck, U. (1999) *World Risk Society* (Cambridge: Polity Press).

Beijing 2008 (2007) *IOC to be Honoured as Champion of the Earth 2007*, http://en.beijing2008.cn/05/61/article214016105.shtml, date accessed 15 September 2008.

Belam, M. (2008) *A Brief History of Olympic Dissent*, http://www.currybet.net (home page), date accessed 25 January 2009.

Bell, M. (2004) *An Invitation to Environmental Sociology*, 2nd edn (Thousand Oaks, CA: Pine Forge Press).

Benton, T. (1991) 'Biology and Social Science', *Sociology*, 25, 1–29.

Benton, T. (1993) *Natural Relations* (London: Verso).

Benton, T. (1994) 'Biology and Social Theory in the Environment Debate', in M. Redclift and T. Benton (eds), *Social Theory and Global Environment* (London: Routledge), pp. 28–50.

Benton, T. (2002) 'Social Theory and Ecological Politics: Reflexive Modernization or Green Socialism' in R. E. Dunlap, F. H. Buttel, P. Dickens, and A. Gijswijt (eds), *Ecological Theory and the Environment* (New York: Oxford; Lanham, Boulder, Rowman & Littlefield), pp. 252–73.

Berger, P. (1986) *The Capitalist Revolution* (New York: Basic Books).

Berger, P. and T. Luckmann (1967) *The Social Construction of Reality* (Garden City: Anchor).

Beriatos, E. and A. Gospodini (2004) '"Glocalising" Urban Landscapes: Athens and the 2004 Olympics', *Cities*, 21/2, 187–202.

Beyer, S. (2006a) 'Environmental Law and Policy in the People's Republic of China', *Chinese Journal of International Law*, 5/1, 185–211.

Beyer, S. (2006b) 'The Green Olympic Movement: Beijing 2008', *Chinese Journal of International Law*, 5/2, 423–40.

Bizios, V. (2011) 'Emission Trading in Greece', *Energy Press* http://www. energypress.gr (home page), date accessed 10 April 2011.

Blowers, A. (1997) 'Environmental Policy: Ecological modernisation or the Risk Society?' *Urban Studies*, 34/5–6, 845–71.

BOCOG (2008) 'Environmental Protection and Meteorology', http://images. beijing2008.cn/upload/lib/bidreport/zt4.pdf, date accessed 10 July 2008.

Börzel, T. A. (2003) *Environmental Leaders and Laggards in Europe: Why There is (Not) A 'Southern Problem'* (Aldershot: Ashgate).

Bookchin, M. (1982) *The Ecology of Freedom* (Palo Alto, CA: Cheshire Books).

Bookchin, M. (1990) *Post–Scarcity Anarchism*, 2nd edn (Montreal and New York: Black Rose Books).

Bookchin, M. (1980/1991) *Toward an Ecological Society* (Montreal and Buffalo: Black Rose Books).

Bookchin, M. (1993) 'What is Social Ecology?' in M. Zimmerman (ed.), *Environmental Philosophy: From Animal Rights to Radical Ecology* (Englewood Cliffs, NJ: Prentice Hall), pp. 354–73.

Botetzagias, I. (2008a) 'The Environmental Impact Assessment and Auditing Process and Greece: Evidence from the Prefectural Level', *Impact Assessment and Project Appraisal*, 26/2, 115–25.

Botetzagias, I. (2008b) 'The Contribution by Classical Sociology' in I. Botetzagias and J. Karamichas (eds) *Environmental Sociology* (Athens: Kritiki), pp. 17–68

Bourdieu, P. (1990) *Other Words* (Cambridge: Polity).

Boykoff, J. (2011) 'The Anti-Olympics', *New Left Review*, 67, 41–59.

Brahm, L. (2010) 'Seeing the Light: China Should Make the Tibetan Plateau Its Linchpin in the Development of Solar Power', *Time*, 176/5, 40.

Brajer, V. and R. W Mead (2003) 'Blue Skies in Beijing? Looking at the Olympic Effect', *The Journal of Environment and Development*, 12/2, 239–63.

Briggs, R., H. McCarthy and A. Zorbas (2004) *16 Days: The Role of the Olympic Truce in the Toolkit for Peace* (London: Demos).

Broudehoux, A. M. (2012) 'Civilizing Beijing: Social Beautification, Civility and Citizenship at the 2008 Olympics' in G. Hayes and J. Karamichas (eds), *Olympic Games, Mega Events and Civil Societies* (Basingstoke: Palgrave Macmillan), pp. 46–67.

Brownell, S. (2006) 'The Beijing Effect', *Olympic Review*, Jun–Sept, 52–5.

Brunet, F. (2009) 'The Economy of the Barcelona Olympic Games' in G. Poynter and I. MacRury (eds) *Olympic Cities: 2012 and the Remaking of London* (Farnham: Ashgate), pp. 97–119.

Brunet, F. and Z. Xinwen (2009) 'The Economy of the Beijing Olympic Games: An Analysis of Prospects and First Impacts' in G. Poynter and I. MacRury (eds) *Olympic Cities: 2012 and the Remaking of London* (Farnham: Ashgate), pp. 163–80.

Bryant, C. (1993) 'Social Self-Organization, Civility and Sociology: A Comment on Kumar's Civil Society', *British Journal of Sociology*, 44/3, 396–401.

Burbank, M. J., G. D. Andranovich and C. H. Heying (2001) *Olympic Dreams: The Impact of Mega–Events on Local Politics* (Boulder, CO: Lynne Rienner Publisher).

Burbank, M. J., G. D. Andranovich and C. H. Heying (2002) 'Mega–Events, Urban Development and Public Policy', *The Review of Policy Research*, 19/3, 179–202.

Business Green Staff (2011) 'Gillard: Australian Carbon Tax Promises "Clean Energy Future"', http://www.businessgreen.com (home page), date accessed 12 July 2011.

Buttel, F. H. (1978) 'Environmental Sociology: A New Paradigm', *The American Sociologist*, 13, 252–6.

Buttel, F. H. (1986) 'Sociology and Environment: The Winding Road Toward Human Ecology', *International Social Science Journal*, 109, 337–56.

Buttel, F. H. (1987) 'New Directions in Environmental Sociology', *Annual Review of Sociology*, 13, 465–88.

Buttel, F. H. (2000a) 'Classical Theory and Contemporary Environmental Sociology: Some Reflections on the Antecedents and Prospects for Reflexive Modernization Theories in the Study of Environment and Society' in G. Spaargaren, A. P. J. Mol and F. H. Buttel (eds) *Environment and Global Modernity* (London: Sage), pp. 17–39.

Buttel, F. H. (2000b) 'Ecological Modernisation as Social Theory', *Geoforum*, 31/1, 57–65.

Buttel, F. H. (2002) 'Environmental Sociology and the Classical Sociological Tradition: Some Observations on Current Controversies' in R. E. Dunlap, F. Buttel, P. Dickens and A. Gijswijt (eds) *Sociological Theory and the Environment* (Boulder, New York, Oxford: Rowman & Littlefield Publishers), pp. 35–50.

Buttel, F. H. (2003) 'Environmental Sociology and the Explanation of Environmental Reform', *Organization and Environment*, 16/3, 306–44.

Buttel, F. H. and W. L. Flinn (1978) 'Social Class and Mass Environmental Beliefs: A Reconsideration', *Environment and Behavior*, 10/3, 433–50.

Buttel, F. H. and C. Humphrey (1987) 'Sociological Theory and the Natural Environment', paper presented at the annual meeting of the American Sociological Association, (Chicago).

Cantelon, H. and M. Letters (2000) 'The Making of the IOC Environmental Policy as the Third Dimension of the Olympic Movement', *International Review for the Sciology of Sport*, 35/3, 294–308.

Capek, S. M. (1993) 'The "Environmental Justice" Frame: A Conceptual Discussion and an Application', *Social Problems*, 40/1, 5–21.

Caratti, P. and L. Ferraguto (2012) 'The Role of Environmental Issues in Mega-Events Planning and Management Processes: Which Factors Count?' in G. Hayes and J. Karamichas (eds) *Olympic Games, Mega Events and Civil Societies* (Basingstoke: Palgrave Macmillan), pp. 109–25.

Carrington, D. (2011) 'Public Belief on Climate Change Weathers Storm, Poll Shows', *The Guardian*, http://www.guardian.co.uk/environment/2011/jan/31/public-belief-climate-change, date accessed 10 February 2011.

Carter, N. (2007) *The Politics of the Environment: Ideas, Activism, Policy*, 2nd edn (Cambridge: Cambridge University Press).

Carter, N. (2009) 'Can the UK Reduce Its Greenhouse Gas Emissions by 2050?' in A. Giddens, S. Latham and R. Liddle (eds) *Building a Low-Carbon Future: The Politics of Climate Change* (London: Policy Network), pp. 111–20.

Cashman, R. (2009) 'Regenerating Sydney's West: Framing and Adapting an Olympic Vision' in G. Poynter and I. MacRury (eds) *Olympic Cities: 2012 and the Remaking of London* (Farnham: Ashgate), pp. 134–44.

Catton Jr., W. R. (2002) 'Has the Durkheim Legacy Mislead Sociology?' in R. E. Dunlap, F. Buttel, P. Dickens and A. Gijswijt (eds) *Sociological Theory and the Environment* (Lanham, Boulder, New York, Oxford: Rowman & Littlefield Publishers), pp. 90–115.

Catton Jr., W. R. and R. Dunlap (1978) 'Environmental Sociology: A New Paradigm', *The American Sociologist*, 13/1, 41–9.

Chalkley, B. and S. Essex (1999a) 'Urban Development through Hosting International Events: A History of the Olympic Games', *Planning Perspectives*, 14, 369–94.

Chalkley, B. and S. Essex (1999b) 'Sydney 2000: The "Green Games?"', *Geography*, 84/4, 299–307.

Christoff, P. (1996) 'Ecological Modernisation, Ecological Modernities', *Environmental Politics*, 5/3, 476–500.

Christoff, P. (2005) 'Policy Autism or Double-Edged Dismissiveness? Australia's Climate Policy under the Howard Government', *Global Change, Peace and Security*, 17/1, 29–44.

Clark, J. (2001) 'Marx's Natures: A Response to Foster and Burkett', *Organization and Environment*, 14/4, 432–42.

Close P., D. Askew and X. Xin (2007) *The Beijing Olympiad: The Political Economy of a Sporting Mega-Event* (London and New York: Routledge).

Cockburn, A. (2011) 'In Fukushima's Wake: How the Greens Learned to Love Nuclear Power', *New Left Review*, 68, 75–9.

Cohen, M. J. (1997) 'Risk Society and Ecological Modernisation: Alternative Visions for Post-Industrial Nations', *Futures*, 29/2, 105–19.

Cohen, M. J. (1998) 'Science and the Environment: Assessing Cultural Capacity for Ecological Modernization', *Public Understanding of Science*, 7/2, 149–67.

Cohen, M. J. (2000) 'Ecological Modernisation, Environmental Knowledge and National Character: A Preliminary Analysis of the Netherlands', in A. P. J. Mol and D. A. Sonnenfeld (eds), *Ecological Modernisation around the*

World: Perspectives and Critical Debates (London and Portland: Frank Cass), pp. 77–106.

COHRE (2007) *Fair Play for Housing Rights: Mega-Events, Olympic Games and Housing Rights* (Geneva: The Centre on Housing Rights and Evictions).

COHRE (2008) *One World, Whose Dream? Housing Rights Violations and the Beijing Olympic Games* (Geneva: The Centre on Housing Rights and Evictions).

Cook, I. G. (2007) 'Beijing 2008' in J. R. Gold and M. M. Gold (eds) *Olympic Cities: City Agendas, Planning, and the World's Games, 1896–2012* (London and New York: Routledge), pp. 286–87.

Coote, A. (2010) *Ten Big Questions about the Big Society and Ten Ways to Make the Best of It*, http://www.neweconomics.org (home page), date accessed 15 July 2010.

Cotgrove, S. and A. Duff (1980) 'Environmentalism, Middle Class Radicalism and Politics', *The Sociological Review*, 28/1, 333–51.

Cotgrove, S. and A. Duff (1981) 'Environmentalism, Values and Social Change', *The British Journal of Sociology*, 32/1, 92–110.

Cotgrove, S. (1982) *Catastrophe or Cornucopia: The Environment, Politics and the Future* (Chichester, Sussex: John Wiley).

Cottrell, M. P. and T. Nelson (2010) 'Not Just the Games? Power, Protest and Politics at the Olympics', *European Journal of International Relations*, ejt. sagepub.com (home page), date accessed 26August 2011.

Crumley, B. (2010) 'Could the Euro's Days Be Numbered?' *Time*, http://www.time.com (home page), date accessed 25 March 2010.

Cuckson, P. (2012) 'London 2012 Olympics: No Water Restrictions for Greenwich Park Equestrian Event', *The Telegraph*, http://www.telegraph.co.uk (home page), date accessed 20 March 2012.

Curran, G. (2009) 'Ecological Modernisation and Climate Change in Australia', *Environmental Politics*, 18/2, 201–17.

Dalton, R., M. Kuechler and W. Bürklin (1990) 'The Challenge of New Movements' in R. Dalton and M. Kuechler (eds), *Challenging the Political Order* (Cambridge: Polity Press), pp. 3–20.

Dave, C. (2005) *The 2012 Bid: Five Cities Chasing the Summer Games* (Bloomington, IN: Author House).

Davis, M. (2002) *Dead Cities: A Natural History* (New York: The New Press).

Defra (2011) *The Natural Choice: Securing the Value of Nature*, http://www.archive.defra.gov.uk/environment/natural/documents/newp-white-paper-110607.pdf, date accessed 15 July 2011.

Della Porta, D., ed. (2007) *The Global Justice Movement: Cross-National and Transnational Parspectives* (Boulder, CO: Paradigm Publishers).

Della Porta, D., M. Andretta, L. Mosca and H. Reiter (2006) *Globalization from Below: Transnational Activists and Global Networks* (Minneapolis, MN: University of Minesota Press).

Dickens, P. (2002) 'A Green Marxism? Labour Processes, Alienation, and the Division of Labour' in R. E. Dunlap, F. Buttel, P. Dickens and A. Gijswijt (eds), *Sociological Theory and the Environment* (Lanham, Boulder, New York, Oxford: Rowman & Littlefield Publishers), pp. 51–72.

Dickens, P. (2004) *Society and Nature* (Cambridge: Polity Press).
Dimopoulos, P., E. Bergmeier and P. Fischer (2006) 'Natura 2000 Habitat Types of Greece Evaluated in the Light of Distribution, Threat and Responsibility', *Proceedings of the Royal Irish Academy*, 106/3, 175–87.
Dobson, A. (1995) *Green Political Thought*, 2nd edn (London and New York: Routledge).
Dodouras, S. and P. James (2006) 'Athens 2004 Olympiad: System Ideas to Map Multidisciplinary Views—Reporting on the Views of the Host Community', *Systemist*, 28/2, 70–81.
Downs, A. (1972), 'Up and Down with Ecology: The "Issue-Attention Cycle", *The Public Interest*, 28, 38–50.
Doyle, T. (2005) *Environmental Movements in Majority and Minority Worlds* (New Brunswick, NJ, and London: Rutgers University Press).
Doyle, T. (2010) 'Surviving the Gang Bang Theory of Nature: The Environmental Movement during the Howard Years', *Social Movement Studies*, 9/2, 155–69.
Dryzek, J. S., D. Downes, C. Hunold, D. Schlosberg and H. K. Hernes (2003) *Green States and Social Movements: Environmentalism in the United States, United Kingdom, and Norway* (Oxford: Oxford University Press).
Dryzek, J. S., C. Hunold, D. Schlosberg, D. Downes and H. K. Hernes (2002) 'Environmental Transformation of the State: The USA, Norway, Germany and the UK', *Political Studies*, 50, 659–82.
Duncan, O. D. (1961) 'From Social System to Ecosystem', *Sociological Inquiry*, 31, 140–9.
Dunlap, R. E (2002a) 'Environmental Sociology: A Personal Perspective on Its First Quarter Century', *Organization & Environment*, 15/1, 10–29.
Dunlap, R. E (2002b) '"Pradigms, Theories and Environmental Sociology' in R. E. Dunlap, F. Buttel, P. Dickens and A. Gijswijt (eds) *Sociological Theory and the Environment*, (Lanham, Boulder, New York, Oxford: Rowman & Littlefield Publishers), pp. 329–50.
Dunlap, R. E. and W. R. Catton, Jr. (1979) 'Environmental Sociology', *Annual Review of Sociology*, 5, 243–273.
Dunlap, R. E. and R. Scarce (1990) 'The Polls: Poll Trends, Environmental Problems and Protection', *Public Opinion Quarterly*, 55, 651–6.
Eatwell, R. (1997) 'Britain' in R. Eatwell (ed.) *European Political Cultures: Conflict or Convergence?* (London and New York: Routledge), pp. 50–68.
Eckersley, R. (1992) *Environmentalism and Political Theory: Toward an Ecocentric Approach* (London: UCL Press).
Ecologic (2011) *Final Report for the Assessment of the 6th Environment Action Programme*, http://ec.europa.eu/environment/newprg/pdf/Ecologic_6EAP_Report.pdf, date accessed 20 April 2011.
Econews.gr (2011a) '"No to Disastrous Investments" by Environmental Organizations', http://www.econews.gr/2011/06/23/ependuseis-perivallon/, date accessed 10 September 2011.
Econews.gr (2011b), 'Letter to G. Papaconstantinou by Environmental Organizations', http://www.econews.gr/2011/07/01/perivallon-mko-papakonstantinou/, date accessed 8 September 2011.

Economou, N. (2010) 'Back on the "Issues Attention Cycle": Labor and the Environment Policy from Hawke to Rudd', http://apsa2010.com.au/full-papers/pdf/APSA2010_0057.pdf, date accessed 10 February 2011.

Economy, E. (2007) 'Environmental Governance: the Emerging Economic Dimension' in N. T. Carter and A. P. J. Mol (eds), *Environmental Governance in China* (London and New York: Routledge), pp. 23–41.

EEA (2008) *Greenhouse Gas Emission Trends and Projections in Europe 2008: Tracking Progress Towards Kyoto Targets* (Copenhagen: European Environment Agency).

European Commission (2009) 'European Attitudes towards Climate Change', *Eurobarometer*, 72.1.

Elafros, Y. (2007) 'As Disaster Looms for the Global Environment, Greece is Still Ruining the Atmosphere with Bad Energy Sources', *Kathimerini*, 10 February.

Elliot, M. and I. Thomas (2009) *Environmental Impact Assessment in Australia: Theory and Practice*, 5th edn (Sydney: The Federation Press).

ESRC (2010) *Olympic Games Impact Study—London 2012 Pre-Games Report*, http://www.uel.ac.uk/geoinformation/documents/UEL_TGIfS_PreGames_OGI_Release.pdf, date accessed 22 January 2010.

Essex, S. and B. Chalkley (1998) 'Olympic Games: Catalysts of Urban Change', *Leisure Studies*, 17/3, 187–206.

Europa (1999) 'Fifth European Community Environment Programme' in *Summaries of EU Legislation*, http://europa.eu/legislation_summaries/other/l28062_en.htm, date accessed 15 February 2007.

European Commission (2009) 'European Attitudes Towards Climate Change', *Eurobarometer*, 72.1.

European Commission (2011) 'Climate Change Report', *Special Eurobarometer*, 372.

Evans, G. (2007) 'London 2012' in J. R. Gold and M. M. Gold (eds) *Olympic Cities: City Agendas, Planning, and the World's Games, 1896–2012* (London and New York: Routledge), pp. 298–317.

Fernández, A. M., N. Font and C. Koutalakis (2010) 'Environmental Governance in Southern Europe: The Domestic Filters of Europeanisation', *Environmental Politics*, 19/4, 557–77.

Finley, M. I. and H. W. Pecket (1976) *The Olympic Games: The First Thousand Years* (Toronto: Clarke, Irwin and Co).

Fischer, J. (1991) 'Beyond Socialism', *New Ecology*, March, 62–5.

Fischer, T. B. (2010) 'Reviewing the Quality of Strategic Environmental Assessment Reports for English Spatial Plan Core Strategies', *Environmental Impact Assessment Review*, 30, 62–9.

Flyvberg, B., N. Bruselius and W. Rothengratter (2003) *Megaprojects and Risk: An Anatomy of Ambition* (Cambridge: Cambridge University Press).

FoE (2011) 'Press Release: Charities Question Cameron's "Greenest Government Ever" Claim One Year On', http://www.foe.co.uk (home page), date accessed 14 May 2011.

Foster, J. B. (1999) 'Marx's Theory of Metabolic Rift: Classical Foundations for Environmental Sociology', *American Journal of Sociology*, 105/2, 366–405.

Foster, J. B. (2002) 'Marx's Ecology in Historical Perspective', *International Socialism Journal*, 96 http://pubs.socialistreviewindex.org.uk/isj96/foster. htm, date accessed 15 September 2007.

Foster, J. B. and P. Burkett (2001) 'Marx and the Dialectic of Organic/Inorganic Relations: A Rejoinder to Salleh and Clark', *Organization & Environment*, 14/4, 451–62.

Frijins, J., Phung Thuy Phuong and A. P. J. Mol (2000) 'Ecological Modernization Theory and Industrialising Economies: The Case of Viet Nam', in A. P. J. Mol and D. A. Sonnenfeld (eds), *Ecological Modernization around the World: Perspectives and Critical Debates* (London: Frank Cass), pp. 257–92.

Furtado, J., T. Belt and R. Jamni, eds (2000) *Economic Development and Environmental Sustainability: Policies and Principles for Durable Equilibrium* (Washington, DC: World Bank).

Fussey, P., J. Coaffee, G. Armstrong, and D. Hobbs, eds (2011) *Securing and Sustaining the Olympic City: Reconfiguring London for 2012 and Beyond* (Farnham: Ashgate).

Games Monitor (n.d.), 'Debunking Olympics Myths', http://www.games-monitor.org.uk/about, date accessed 21 March 2012.

Garcia, B. (2007) 'Sydney 2000', in J. R. Gold and M. M. Gold (eds), *Olympic Cities: City Agendas, Planning, and the World's Games, 1896–2012* (London and New York: Routledge), pp. 235–64.

George, A. L. and T. J. McKeown (1985) 'Case Studies and Theories of Organizational Decision Making', *Advances in Information Processing in Organizations*, 2, 21–58.

Georgi, N. (2010) 'Public Space in Ancient Greece' in E. I. Manolas (ed.) *Natural Environment in Ancient Greece* (Orestiada: Department of Forestry and Management of Natural Resources), pp. 85–94.

Gersmann, H. (2011) 'UK's Faith in Nuclear Power Threatens Renewable, says German Energy Expert', *The Guardian*, http://www.guardian.co.uk (home page), date accessed 12 December 2011.

G-ForSE (n.d.) 'The Environment and the Olympic Movement', http://www.g-forse.com/enviro/Olympic.html, date accessed 20 November 2009.

Giddens, A. (1990) *The Consequences of Modernity* (Cambridge: Polity Press).

Giddens, A. (1991) *Modernity and Self-Identity: Self and Society in the Late Modern Age* (Cambridge: Polity Press).

Giddens, A. (2008) *Sociology*, 6th edn (Cambridge: Polity Press).

Giddens, A. (2009) *The Politics of Climate Change* (Cambridge: Polity Press).

Girginov, V. and L. Hills (2010) 'A Sustainable Sports Legacy: Creating a Link Between the London Olympics and Sports Participation' in V. Girginov (ed.), *The Olympics: A Critical Reader* (Abingdon and New York: Routledge), pp. 430–47.

Gold, M. M. (2007) 'Athens 2004' in J. R. Gold and M. M. Gold (eds) *Olympic Cities: City Agendas, Planning, and the World's Games, 1896–2012* (London and New York: Routledge), pp. 265–85.

Goldblatt, D. (1996) *Social Theory and the Environment* (Cambridge: Polity Press).

Goldenberg, S. (2009) 'Global Poll Finds 73% Want Higher Priority for Climate Change', *The Guardian*, http://www.guardian.co.uk (home page), date accessed 30 September 2009.

Goldsmith, E., R. Allen, M. Allaby, J. Davoll and S. Lawrence (1972) 'A Blueprint for Survival', *The Ecologist*, 2/1, January, 1–43.

Gouta, A. (2009) 'Ancient Greeks and the Environment: Laws, Myths and Thousands of Years of Ecological "Sensitivities"', http://portal.tee.gr/portal/page/portal/teetkm/, date accessed 30 November 2010.

Green Games Watch 2000 (1999) *Environmental Compliance of Selected Olympic Venues*, (Bondi Junction, New South Wales: Green Games Watch 2000).

Greenpeace (2004a) *How Green the Games? A Greenpeace Assessment of the Environmental Performance of the Athens 2004 Olympics* (Athens: Greenpeace).

Greenpeace (2004b) *Olympic Games—Athens 2004. Promises: Always Green, Always Forgotten. An Assessment of the Environmental Dimension of the Games.* (Athens: Greenpeace.)

Gross, M. (2000) 'Classical Sociology and the Restoration of Nature: The Relevance of Émile Durkheim and Georg Simmel', *Organization & Environment*, 13/3, 277–91.

Guardian, The (2011) 'Indices of Multiple Deprivation: Find the Poorest Place in England', http://www.guardian.co.uk (home page), date accessed 15 May 2011.

Gursoy, D. and K. W. Kendall (2006) 'Hosting Mega Events, Modeling Locals' Support', *Annals of Tourism Research*, 35/3, 603–23.

Habermas, J. (1989) *The Theory of Communicative Action* (Cambridge: Polity Press).

Hajer, M. A. (1995) *The Politics of Environmental Discourse: Ecological Modernisation and the Policy Process* (Oxford: Oxford University Press).

Hajer, M. A. (1996) 'Ecological Modernisations as Cultural Politics' in S. Lash, B. Szerszynski and B. Wynne (eds) *Risk, Environment and Modernity: Towards a New Ecology* (Thousand Oaks, New Delhi: Sage), pp. 246–268.

Hall, C. M. (2001) 'Imaging, Tourism and Sports Event Fever: The Sydney Olympics and the Need for a Social Charter for Mega-Events', in C. Gratton and I. P. Henry (eds) *Sport in the City: The Role of Sport in Economic and Social Regeneration* (London: Routledge) pp. 166–83.

Hall, C. M. and J. Hodges (1996) 'The Party's Great, but What about the Hangover? The Housing and Social Impacts of Mega-Events with Special Reference to the 2000 Sydney Olympics', *Festival Management & Event Tourism*, 4, 13–20.

Hannigan, J. (1995) *Environmental Sociology: A Social Constructionist Perspective* (London and New York: Routledge).

Hannigan, J. (2006) *Environmental Sociology*, 2nd edn (London and New York: Routledge).

Harmel, R. and K. Janda (1994) 'An Integrated Theory of Party Goals and Party Change', *Journal of Theoretical Politics*, 6/3, 259–87.

Harris, P. J., E. Harris, S. Thompson, B. Harrris-Roxas and L. Kemp (2009) 'Human Health and Wellbeing in Environmental Impact Assessment in

New South Wales, Australia: Auditing Health Impacts within Environmental Assessment of Major Projects', *Environmental Impact Assessment Review*, 29/5, 310–18.

Harvey, F. (2011) 'Worst Ever CO2 Emissions Leave Climate on the Brink', *The Guardian*, 30 May 2011, 1–2.

Hayes, G. and J. Horne (2011) 'Sustainable Development, Shock and Awe? London 2012 and Civil Society', *Sociology*, 45/5, 749–64.

Hayes, G. and J. Karamichas (2012a) 'Introduction: Sports Mega-Events, Sustainable Development, and Civil Societies' in G. Hayes and J. Karamichas (eds) *Olympic Games, Mega-Events and Civil Societies* (Basingstoke: Palgrave Macmillan), pp. 1–39.

Hayes, G. and J. Karamichas (2012b) 'Conclusion: Sports Mega-Events: Disputed Places, Systemic Contradictions, and Critical Moments' in G. Hayes and J. Karamichas (eds), *Olympic Games, Mega-Events and Civil Societies* (Basingstoke: Palgrave Macmillan), pp. 249–61.

Hiller, H. H. (2000) 'Toward an Urban Sociology of Mega-Events', *Research in Urban Sociology*, 5, 181–205.

Hilton, M., J. MacKay, N. Crowson and J. F. Mouhot (2010) ' "The Big Society": Civic Participation and the State in Modern Britain', *History & Policy http://www.historyandpolicy.org/papers/policy-paper-103.html*, date accessed 12 March 2011.

Hoberman, J. (2004) 'Sportive Nationalism and Globalization' in J. Bale and M. K. Christensen (eds) *Post-Olympism? Questioning Sport in the Twenty-First Century* (Oxford: Berg), pp. 177–88.

Hogan, D. J. (2000) 'Socio-Demographic Dimensions of Sustainability: Brazilian Perspectives', www.ciesin.columbia.edu/repository/pern/papers/ISARio2000.doc, date accessed 11 April 2011.

Horne, J. (2007) 'The Four "Knowns" of Sports Mega-Events', *Leisure Studies*, 26/1, 81–96.

Horne, J. (2012) 'The Four 'Cs' of Sports Mega-Events: Capitalism, Connections, Citizenship, and Contradictions' in G. Hayes and J. Karamichas (eds), *Olympic Games, Mega Events and Civil Societies* (Basingstoke: Palgrave Macmillan), pp. 31–45.

Horne, J. and W. Manzenreiter (2006) 'An Introduction to the Sociology of Sports Mega-events' in J. Horne and W. Manzenreiter (eds), *Sports Mega-Events: Social Scientific Analyses of a Global Phenomenon* (Oxford: Blackwell), pp. 1–24.

Huber, J. (2000) 'Towards Industrial Ecology: Sustainable Development as a Concept of Ecological Modernization', *Journal of Environmental Policy & Planning*, 2, 269–85.

Huber, M. (2003) 'Review of Megaprojects and Risk', *British Journal of Sociology*, 54/4, 593–608.

Inglehart, R. (1971) 'The Silent Revolution in Europe: Intergenerational Change in Post-Industrial Societies', *American Political Science*, 65, 991–1017.

Inglehart, R. (1977) *The Silent Revolution: Changing Values and Political Styles Among Western Publics* (Princeton, NJ: Princeton University Press).

Inglehart, R. (1990) *Culture Shift in Advanced Industrial Society* (Princeton, NJ: Princeton University Press).

IOC (1996) *Manual for Candidate Cities for the Games of the XXVIII Olympiad 2004* (Lausanne: International Olympic Committee).

IOC (2000) *Beijing 2008, Candidate City* (Lausanne: International Olympic Committee).

IOC (2004), *2012 Candidature Procedure and Questionnaires. Games of the XXX Olympiad in 2012* (Lausanne: International Olympic Committee).

IOC (2005) *Report of the IOC Evaluation Commission for the Games of the XXX Olympiad in 2012* (Lausanne: International Olympic Committee).

IOC (2007) *Olympic Charter* (Lausane: International Olympic Committee).

IOC (2011) *Olympic Charter* (Lausane: International Olympic Committee).

Iles, A. (2007) 'Of Lammas Land and Olympic Dreams', *Games Monitor*, http://www.gamesmonitor.org.uk/node/322, date accessed 15 January 2010.

Irwin, A. (2001) *Sociology and the Environment: A critical Introduction to Society, Nature and Knowledge* (Cambridge: Polity Press).

Jamieson, A. (2009) 'Beijing Olympics Were the Most Polluted Games Ever, Researchers Say', *The Telegraph*, http://www.telegraph.co.uk (home page), date accessed 30 June 2009.

Jänicke, M. and H. Weidner, eds (1997) *National Environmental Policies: A Comparative Study of Capacity-Building (13 Countries)* (New York/Berlin: Springer-Verlag).

Jarvie, G., D. J. Hwang and M. Brennan (2008) *Sport, Revolution and the Beijing Olympics* (Oxford and New York: Berg).

Järvikowski, T. (1996) 'The Relation of Nature and Society in Marx and Durkheim', *Acta Sociologica*, 39/1, 73–86.

Jehlicka, P. (1992) 'Environmentalism in Europe', paper presented to the British Sociological Association Conference, April.

Johnson, D. (2004) 'Ecological Modernization, Globalization and European-ization: A Mutually Reinforcing Nexus?' in J. Barry, B. Baxter and R. Dunphy (eds) *Europe, Globalization and Sustainable Development* (London and New York: Routledge), pp. 152–67.

Kalogeratou, A. (2007) 'Did the Olympic Games Take Place in 2004?' in T. Doulkeri (ed.), *Sports, Society and Mass Media: The Case of the Olympic Games of Athens 2004* (Athens: Papazisi), pp. 159–79.

Karamichas, J. (2003) 'Civil Society and the Environmental Problematic: A Preliminary Investigation of the Greek and Spanish Cases', paper presented at the ECPR joint sessions of workshops, University of Edinburgh, 28 March.

Karamichas, J. (2005) 'Risk versus National Pride: Conflicting Discourses over the Construction of a High Voltage Power Station in the Athens Metropolitan Area for Demands of the 2004 Olympics', *Human Ecology Review*, 12/2, 133–42.

Karamichas, J. (2007a) 'Key Issues in the Study of New and Alternative Social Movements in Spain: The Left, Identity and Globalizing Processes', *South European Society & Politics*, 12/3, 273–94.

Karamichas, J. (2007b) 'The Impact of the Summer 2007 Forest Fires in Greece: Recent Environmental Mobilizations, Cyber-Activism and Electoral Performance', *South European Society & Politics*, 12/4, 521–34.

Karamichas, J. (2008) 'Red and Green Facets of Political Ecology: Accounting for Electoral Prospects in Greece', *Journal of Modern Greek Studies*, 26, 311–36.

Karamichas, J. (2012a) 'Olympic Games as an Opportunity for the Ecological Modernisation of the Host Nation: The Cases of Sydney 2000 and Athens 2004' in G. Hayes and J. Karamichas (eds) *Olympic Games, Mega–Events and Civil Societies* (Basingstoke: Palgrave Macmillan), pp. 151–71.

Karamichas, J. (2012b), 'The Olympics and the Environment' in S. H. J. Lenskyj and S. Wagg (eds) *Handbook of Olympic Studies* (Basingstoke: Palgrave Macmillan), pp. 249–61.

Karamichas, J. (2012c) 'A Source of Crisis? Assessing Athens 2004' in S. H. J. Lenskyj and S. Wagg (eds) *Handbook of Olympic Studies* (Basingstoke: Palgrave Macmillan), pp. 163–77.

Katsikas, C. and E. Nikolaidou (2003) *The Olympic Games in Ancient Greece: Unseen Facets* (Athens: Savvalas).

Katz, R. and P. Mair (1995) 'Changing Models of Party Organization and Party Democracy: The Emergence of the Cartel Party' *Party Politics*, 1/1, 5–21.

Kavoulakos, K. I. (2008) 'Protestation and Claiming of Public Places: A Movement in the City of Athens in the 21st Century' in D. Emanouel, E. Zakopoulou, R. Kautantzoglou, T. Maloutas and A. Hadjiyanni (eds) *Social and Spatial Transformationsin 21st Century Athens* (Athens: EKKE).

Kazantzopoulos, G. (2002) 'The Environmental Strategy for the Olympic Projects', in Hellenic Environmental Law Society (ed.), *Olympic Games and the Environment – Conference Proceedings Athens 1-2/02/2001* (Athens and Komotini: Sakkoulas), pp. 109–12.

Kidd, B. (2005) 'Another World is Possible: Recapturing Alternative Olympic Histories, Imagining Different Games' in K. Young and K. B. Wamsley (eds), *Global Olympics: Historical and Sociological Studies of the Modern Games* (Oxford: Elsevier), pp. 143–58.

Kirchheimer, O. (1966) 'The Transformation of the Western European Party Systems' in J. LaPalombara and M. Weiner (eds), *Political Parties and Political Development*, (Princeton: Princeton University Press), pp. 177–200.

Kitschelt, H. (1988) 'Left-Libertarian Parties: Explaining Innovation in Competitive Party Systems', *World Politics*, 40/2, 194–234.

Kitschelt, H. (1990) 'New Social Movements and the Decline of Party Organization' in R. Dalton and M. Kuechler (eds) *Challenging the Political Order* (Cambridge: Polity Press), pp. 179–208.

Kitschelt, H. and S. Hellemans (1990) 'The Left-Right Semantics and the New Politics Cleavage', *Comparative Political Studies*, 23/2, 210–38.

Klein, N. (2007) *The Shock Doctrine: The Rise of Disaster Capitalism* (London: Allen Lane/Penguin).

Koutalakis, C. (2004) 'Environmental Compliance in Italy and Greece: The Role of Non-State Actors', *Environmental Politics*, 13/4, 754–74.

Kriesi, H. (1989) 'New Social Movements and the New Class in the Netherlands', *American Journal of Sociology*, 94/5, 1078–116.

Kritikopoulou, D. (2007) 'The Olympic Ideal: A Historical Overview of the Olympic Games Institution through the Philosophical Theorisation of the Values of Olympism' in T. Doulkeri (ed.) *Sports, Society and Mass Media. The Case of the Olympic Games of Athens 2004* (Athens: Papazisi), pp. 47–75.

Krüger, A. (2004) 'Was the 1936 Olympics the First Postmodern Spectacle?' in J. Bale and M. K. Christensen (eds) *Post-Olympism? Questioning Sport in the Twenty-First Century* (Oxford: Berg), pp. 35–50.

Krüger, A. (2005) 'The Nazi Olympics of 1936', in K. Young and K. B. Wamsley (eds), *Global Olympics: Historical and Sociological Studies of the Modern Games* (London: Elsevier), pp. 43–58.

Kumar, K. (1994) 'Civil society again: a reply to Christopher Bryant's "Social self-organization, civility and sociology"', *British Journal of Sociology*, 45/1, 127–31.

La Spina, A. and G. Sciortino (1993) 'Common Agenda, Southern Rules: European Integration and Environmental Change in the Mediterranean States', in J. D. Liefferink, P. D. Lowe and A. P. J. Mol (eds), *European Integration and Environmental Policy* (London and New York: Belhaven), pp. 99–113.

Laver, M. and K. A. Shepsle (1999) 'How Political Parties Emerged from the Primeval Slime: Party Cohesion, Party Discipline, and the Formation of Governments' in S. Bowler, D. M. Farrell and R. S. Katz (eds) *Party Discipline and Parliamentary Government* (Columbus: Ohio State University Press), pp. 23–48.

Lee, Y. and A. So (1999) *Asia's Environmental Movements* (Armonk, NY: M.E. Sharpe).

Lenskyj, H. J. (2000) *Inside the Olympic Industry: Power, Politics and Activism* (Albany, NY: State University of New York Press).

Lenskyj, H. J. (2002) *The Best Olympics Ever? Social Impacts of Sydney 2000* (Albany, NY: State University of New York Press).

Lenskyj, H. J. (2008) *Olympic Industry Resistance: Challenging Olympic Power and Propaganda* (Albany, NY: State University of New York Press).

Leonard, H. J. (1988) *Pollution and the Struggle for the World Product: Multinational Corporations, Environment and International Comparative Advantage* (Cambridge: Cambridge University Press).

Lesjø, J. H. (2000) 'Lillehammer 1994', *International Review for the Sciology of Sport*, 35/3, 282–93.

Li, L. M., A. J. Dray-Novey and H. Kong (2007) *Beijing: From Imperial Capital to Olympic City* (Basingstoke: Palgrave Macmillan).

Lialios, G. (2011) 'What Did the First Minister of the Environment Left to the Environment', http://www.kathimerini.gr/4dcgi/_w_articles_oiko1_2_01/09/2011_1296107, date accessed 3 November 2011.

Lichtheim, G. (1964) *Marxism: An Historical and Critical Study* (New York: Routledge).

Lin, S., D. Zhao and D. Marinova (2009) 'Analysis of the Environmental Impact of China Based on STIRPAT Model', *Environmental Impact Assessment Review*, 29, 341–47.

Lo, A. Y. (2010) 'Active Conflict or Passive Coherence? The Political Economy of Climate Change in China', *Environmental Politics*, 19/6, 1012–17.

Lober, D. J. (1993) 'Beyond Self-Interest: A Model of Public Attitudes Towards Waste Facility Siting', *Journal of Environmental Planning and Management*, 36/3, 345–63.

Lober, D. J. (1995) 'Why Protest? Public Behavioral and Attitudinal Response to Siting a Waste Disposal Facility', *Policy Studies Journal*, 23/3, 499–518.

Logan, J. and Molotch, H. (1987) *Urban Fortunes: The Political Economy of Place* (Berkeley: University of California Press).

Loh, C. (2008) 'Clearing the Air' in M. Worden (ed.) *China's Great Leap: The Beijing Games and Olympian Human Rights Challenges* (New York, London, Melbourne, Toronto: Seven Stories Press), pp. 235–45.

London 2012 (2004) 'Response to the Questionnaire for Cities Applying to Become Candidate Cities to Host the Games of the XXX Olympiad and the Paralympic Games in 2012', http://www.london2012.com/documents/candidate-files/questionnaire-response-english.pdf, date accessed 02 February 2009.

London 2012 (2007) 'Sustainability', http://www.london2012.com/about-us/sustainability/, date accessed 15 July 2007.

London 2012 (2008a) 'Towards a One Planet 2012 Sustainability Plan Update', http://www.london2012.com/documents/locog-publications/sustainability-plan-december-08.pdf, date accessed 02 February 2009.

London 2012 (2008b) 'Sustainability Plan Progress Report Card', http://www.london2012.com/documents/locog-publications/sustainability-report-card-december-08.pdf, date accessed 02 February 2009.

Lourdis, G. (2009) 'The Olympic Village of Athens 2004 (the Greatest Olympic Project)', http://library.tee.gr/digital/m2482/m2482_contents.htm, date accessed 12 April 2010.

Lovelock, J. (1979) *Gaia* (Oxford: Oxford University Press).

Lovelock, J. (2006) *The Revenge of Gaia* (London: Penguin).

Luscombe, D. (1998) 'Promises' in R. Cashman and A. Hughes (eds) *The Green Games: A Golden Opportunity* (Sydney: University of New South Wales, Centre for Olympic Studies), pp. 14–17.

MacKenzie, J. D. (2006) *Moving Towards Sustainability in the Olympic Games Planning Process* (Burnaby, Canada: Simon Fraser University Library).

MacRury, I. and G. Poynter (2009) 'Olympic Cities and Social Change' in G. Poynter and I. MacRury (eds) *Olympic Cities: 2012 and the Remaking of London* (Farnham: Ashgate), pp. 303–26.

Malfas, M., E. Theodoraki and B. Houlihan (2004) 'Impacts of the Olympic Games as Mega-Events', *Municipal Engineer*, 157/3, 209–20.

Malkoutzis, N. (2011) *Greece–A Year in Crisis: Examining the Social and Political Impact of an Unprecedented Austerity Programme* (Berlin: Friedrich-Ebert-Stiftung).

Marks, D. (2010) 'China's Climate Change Policy Process: Improved but still Weak and Fragmented', *Journal of Contemporary China*, 19/67, 971–86.

Marquart-Pyatt, S. T. (2007) 'Concern for the Environment among General Publics: A Cross-National Study, *Society and Natural Resources*, 20/10, 833–98.

Marshall, P. (2008) *Demanding the Impossible: A History of Anarchism* (London: Harper Collins).

Martell, L. (1994) *Ecology and Society: An Introduction* (Cambridge: Polity Press).

Martens, S. (2007) 'Public Participation with Chinese Characteristics: Citizen Consumers in China's Environmental Management' in N. T. Carter and A. P. J. Mol (eds) *Environmental Governance in China* (London and New York: Routledge), pp. 63–82.

Marx, K. (1976 [1867]) *Capital*, vol. 1 (New York: Vintage).

Maslow, A. H. (1970) *Motivation and Personality* (New York: Harper & Row).

McAdam, D. (1996) 'Conceptual Origins, Current Problems, Future Directions' in D. McAdam, J. D. McCarthy and M. N. Zald (eds) *Comparative Perspectives on Social Movements* (Cambridge: Cambridge University Press), pp. 24–40.

McGeoch, R. (1999) 'The Green Games: The Legal Obligations that have Arisen from the "Green" Bid', *The University of New South Wales Law Journal*, 22/3, 708–20.

Meadows, D., J. Randers and W. Behrens (1972) *The Limits to Growth* (London: Earth Island).

Melucci, A. (1980) 'The New Social Movements: A Theoretical Approach', *Social Science Information*, 19, 199–226.

Melucci, A. (1985) 'The Symbolic Challenge of Contemporary Movements', *Social Research*, 52/4, 789–815.

Melucci, A. (1988) 'Social Movements and the Democratization of Everyday Life' in J. Keane (ed.) *Civil Society and the State* (London: Verso), pp. 247–59.

Mol, A. P. J. (1995) *The Refinement of Production: Ecological Modernization Theory and the Chemical Industry* (Utrecht, Netherland: Van Arkel).

Mol, A. P. J. (1996) 'Ecological Modernisation and Institutional Reflexivity: Environmental Reform in the Late Modern Age', *Environmental Politics*, 5/2, 302–23.

Mol, A. P. J. (2000) 'The Environmental Movement in an Era of Ecological Modernisation', *Geoforum*, 32, 45–56.

Mol, A. P. J. (2001) *Globalization and Environmental Reform: Ecological Modernization Theory and the Global Economy* (Cambridge, MA: MIT Press).

Mol, A. P. J. (2010) 'Sustainability as Global Attractor: The Greening of the 2008 Beijing Olympics', *Global Networks*, 10/4, 510–28.

Mol, A. P. J. and D. A. Sonnenfeld (2000) 'Ecological Modernisation around the World: An Introduction', in A. P. J. Mol and D. A. Sonnenfeld (eds), *Ecological Modernisation around the World: Perspectives and Critical Debates* (London and Portland: Frank Cass), pp. 3–14.

Mol, A. P. J. and G. Spaargaren (1993) 'Environment, Modernity and the Risk Society: The Apocalyptic Horizon of Environmental Reform', *International Sociology*, 8/4, 431–459.

Mol, A. P. J. and L. Zhang (2012) 'Sustainability as Global Norm: The Greening of Mega-Events in China' in G. Hayes and J. Karamichas (eds) *Olympic Games, Mega-Events and Civil Societies* (Basingstoke: Palgrave Macmillan), pp. 126–50.

Molotch, H. (1976) 'The City as a Growth Machine: Toward a Political Economy of Place', *American Journal of Sociology*, 82/2, 309–32.

Monbiot, G. (2011a) 'Japan Nuclear Crisis Should Not Carry Weight in Atomic Energy Debate', *The Guardian*, http://www.guardian.co.uk (home page), date accessed 30 March 2011.

Monbiot, G. (2011b) 'Why Fukushima Made Me Stop Worrying and Love Nuclear Power', *The Guardian*, http://www.guardian.co.uk (home page), date accessed 30 March 2011.

Monbiot, G. (2011c) 'Cameron's "Green Growth" Policy Looks Naive Today: It Will Look Cynical in 2007', *The Guardian*, 14 May, p. 27.

Morrison, D. E., K. Hornback and W. K. Warner (1972) 'The Environmental Movement: Some Preliminary Observations and Predictions' in W. R. Burch Jr., N. H. Cheek Jr. and L. Taylor (eds) *Social Behaviour, National Resources and the Environment* (New York: Harper & Row), pp. 259–79.

Müller-Rommel, F. (1990) 'New Political Movements and "New Politics" Parties in Western Europe' in R. Dalton and M. Kuechler (eds) *Challenging the Political Order* (Cambridge: Polity Press), pp. 209–31.

Mumford, L. (1966) *Technics and Human Development: The Myth of the Machine*, Vol. 1 (London: HBJ Publishers).

Murphy, R. (1995) 'Sociology as If Nature Did Not Matter: An Ecological Critique', *The British Journal of Sociology*, 46/4, 688–707.

Murphy, R. (2002) 'Ecological Materialism and the Sociology of Max Weber' in R. E. Dunlap, F. Buttel, P. Dickens and A. Gijswijt (eds) *Sociological Theory and the Environment* (Lanham, Boulder, New York, Oxford: Rowman & Littlefield Publishers), pp. 73–89.

Mythen, G. (2004) *Ulrich Beck: A Critical Introduction to the Risk Society* (London: Pluto Press).

No to London 2012 (n.d), 'Say No to London 2012', http://www.nolondon 2012.org/, date accessed 15 March 2009.

O'Connor, J. (1998) *Natural Causes: Essays in Ecological Marxism* (New York: The Guilford Press).

Odysseas, (2011) '10 ENGOs Denounce the Plundering of the "Green Fund", http://ecology-salonika.org/2010/10/20/plunder-of-the-green-fund-reported-by-10-environmental-organizations/, date accessed 22 October 2011.

Offe, C. (1985) 'New Social Movements: Challenging the Boundaries of Institutional Politics', *Social Research*, 52, 817–68.

Olds, K. (1998) 'Urban Mega-Events, Evictions and Housing Rights: The Canadian Case', *Current Issues in Tourism*, 1/1, 2–46.

Opschoor, J. B. (1997) 'Industrial Metabolism, Economic Growth and Institutional Change' in M. Redclift and G. Woodgate (eds) *The International Handbook of Environmental Sociology* (Northampton, MA: Edward Elgar), pp. 274–86.

Pagoulatos, G. (2010) *The Greek Economy and the Potential for Green Development* (Berlin: Friedrich-Ebert-Stiftung).

Pakulski, J., B. Tranter and S. Crook (1998) ' Synamics of Environmental Issues in Australia: Concerns, Clusters and Carriers', *Australian Journal of Political Science*, 33/2, 235–53.

Panagiotopoulou, R. (2009) 'The 28th Olympic Games in Athens 2004' in G. Poynter and I. MacRury (eds), *Olympic Cities: 2012 and the Remaking of London* (Farnham: Ashgate), pp. 145–62.

Papadakis, E. (1993) *Politics and the Environment* (St. Leonards, NSW: Allen ans Unwin).

Papadakis, E. (2002) 'Environmental Capacity Building in Australia' in H. Weidner and M. Jänicke (eds) *Capacity Building in National Environmental Policy* (London: Springer), pp. 19–44.

Papadopoulos, A. G. and C. Liarikos (2007) 'Dissecting Changing Rural Development Networks: The Case of Greece', *Environment and Planning C: Government and Policy*, 25, 292–313.

Payne, M. (2006) *Olympic Turnaround: How the Olympic Games Stepped Back from the Brink of Wxtinction to Become the World's Best Known Brand* (London: Praeger).

Pellow, D. N. (2000) 'Environmental Inequality Formation: Toward a Theory of Environmental Injustice', *American Behavioral Scientist*, 43/4, 581–601.

Pellow, D. N. (2004) 'The Politics of Illegal Dumping: An Environmental Justice Framework', *Qualitative Sociology*, 27/4, 511–25.

Perrottet, T. (2004) *The Naked Olympics: The True Story of the Ancient Games* (New York: Random House).

Pietsch, J. and I. McAllister (2010) '"A Diabolical Challenge": Public Opinion and Climate Change Policy in Australia', *Environmental Politics*, 19/2, 217–36.

Poguntke, T. (1987) 'New Politics and Party Systems: The Emergence of a New Type of Party', *West European Politics*, 10, 76–88.

Porritt, J. (2011) *"The Greenest Government Ever": One Year On, A Report to Friends of the Earth*, http://www.foe.co.uk/resource/reports/greenest_gvt_ever.pdf, date accessed 10 June 2011.

Porter, M. and C. van der Linde (1995) 'Green and Competitive-Ending the Stalemate', *Harvard Business Review*, 73, 120–34.

Pound, R. (2003) *Olympic Games Study Commission: Report to the 115th IOC Session, Prague, July 2003* (Lausanne: International Olympic Committee).

Pound, R. (2004) *Inside the Olympics: A Behind the Scenes Look at the Politics, the Scandals and the Glory of the Games* (Chiechester: Wiley).

Pound, R. (2008) 'Olympian Changes: Seoul and Beijing' in M. Worden (ed.) *China's Great Leap: The Beijing Games and Olympian Human Rights Challenges* (New York, London, Melbourne, Toronto: Seven Stories Press), pp. 85–97.

Poynter, G. (2009) 'London: Preparing for 2012' in G. Poynter and I. MacRury (eds), *Olympic Cities: 2012 and the Remaking of London* (Farnham: Ashgate), pp. 183–99.

Preuss, H. (2004) *The Economic of Staging the Olympics: A Comparison of the Games 1972–2008* (Cheltenham: Edward Elgar).

Pridham, G. and M. Cini (1994) 'Environmental Standards in the European Union: Is There a Southern Problem?' in M. Faure, J. Vervaele and

A. Weale (eds), *Environmental Standards in the EU in an Interdisciplinary Framework* (Antwerp: Maklu), pp. 251–77.

Psaropoulos, J. (2008) 'Greece's Suspension from Kyoto', *Athens News*, 28 June.

Renou, X. (2012) 'Resisting the Torch' in G. Hayes and J. Karamichas (eds), *Olympic Games, Mega-Events and Civil Societies* (Basingstoke: Palgrave Macmillan), pp. 236–46.

Richerzhagen, C. and I. Scholz (2008) 'China's Capacities for Mitigating Climate Change', *World Development*, 36/2, 308–24.

Robertson, R. (1992) *Globalization, Social Theory and Social Culture* (London: Sage).

Roche, M. (2000) *Mega-Events and Modernity: Olympics and Expos in the Growth of Global Culture* (London and NY: Routledge).

Roche, M. (2006) 'Mega-Events and Modernity Revisited: Globalization and the Case of the Olympics' in J. Horne and W. Manzenreiter (eds) *Sports Mega-Events. Social Scientific Analyses of a Global Phenomenon* (Oxford: Blackwell), pp. 27–40.

Rootes, C. (2008) 'The First Climate Change Elections? The Australian General Election of 24 November 2007', *Environmental Politics*, 17/3, 473–80.

Rootes, C. (2009) 'Environmental Protests, Local Campaigns and the Environmental Movement in England' Paper prepared for presentation to Workshop 5:'*Professionalization and Individualized Collective Action: Analyzing New 'Participatory' Dimensions in Civil Society*' European Consortium for Political Research Joint Sessions, Lisbon 14–19 April, 2009.

Rootes, C. (2011) 'Denied, Deferred, Triumphant? Climate Change, Carbon Trading and the Greens in the Australian Federal Election of 21 August 2010' *Environmental Politics*, 20/3, 410–7.

Rucht, D. (1990) 'The Strategies and Action Repertoires of New Movements' in R. Dalton and M. Kuechler (eds) *Challenging the Political Order* (Cambridge: Polity Press), pp. 156–75.

Rustin, S. (2011) 'Let the Games Begin', *The Guardian*, 23 July, 40.

Rutheiser, C. (1996) *Imagineering Atlanta. The Politics of Place in the City of Dreams.* (London and New York: Verso).

Rutheiser, C. (1999) 'Making Place in the Nonplace Urban Realm: Notes on the Revitalization of Downtown Atlanta' in S. M. Low (ed.) *Theorizing the City: The New Urban Anthropology Reader* (New Brunswick, NJ, and London: Rutgers University Press), pp. 317–41.

Sadd, D. and I. Jones (2009) 'Long-Term Legacy Implications for Olympic Games' in R. Raj and J. Musgrave (eds), *Event Management and Sustainability* (Wallingford: CABI Publishing), pp. 90–8.

Sassen, S. (2007) *A Sociology of Globalization* (New York: W. W. Norton & Co).

Schnaiberg, A. (1980) *Environment: From Surplus to Scarcity* (New York: Oxford University Press).

Schreurs, M. A. and Y. Tiberghien (2007) 'Multi-Level Reinforcement: Explaining European Union Leadership in Climate Change Mitigation', *Global Environmental Politics*, 7/4, 19–46.

Schumacher, E. F. 1988 [1973]) *Small is Beautiful: A Study of Economics as if People Mattered* (London: Abacus).

Scott, A. (1990) *Ideology and the New Social Movements* (London: Unwin Hyman).

Short, J. (2003) 'Going for gold: globalizing the Olympics, localizing the Games', *Globalization and World Cities (GaWC) Research Bulletin*, 10.

Sklair, L. (1994) 'Global Sociology and Global Environmental Change' in M. Redclift and T. Benton (eds) *Social Theory and the Global Environment* (London: Routledge), pp. 205–27.

Smales, N. (2008) Presentation on Olympic Burroughs given at the international conference, *Mega Events and Civil Societies*, at Queen Mary, University of London, 26–27 June 2008.

So, A. Y. (1990) *Social Change and Development: Modernization, Dependency and World – System Theories* (London: Sage).

SOCOG (1993) *Sydney Olympics 2000 Bid* (Sydney: Sydney Organising Committee for the Olympic Games).

Spaargaren, G. (2000) 'Ecological Modernisation Theory and the Changing Discourse of Environment and Modernity' in G. Spaargaren, A. P. J. Mol and F. Buttel (eds) *Environment and Global Modernity* (London, Thousands Oaks & New Delhi: Sage), pp. 41–72.

Spencer, R. (2008) 'Beijing Olympics: "Ethnic" Children Exposed as Fakes in Opening Ceremony', *The Telegraph*, http://www.telegraph.co.uk (home page), date accessed 20 September 2008.

Spilling, O. R. (1996) 'Mega-Event as Strategy for Regional Development: The Case of the 1994 Lillehammer Winter Olympics', *Entrepreneurship & Regional Development*, 8/4, 321–44.

Spivey, N. (2004) *The Ancient Olympics* (Oxford: Oxford University Press).

Spowers, R. (2002) *Rising Tides: The History and Future of the Environmental Movement* (Edinburgh: Canongate).

Stern, P. C. (2000) 'New Environmental Theories: Toward a Coherent Theory of Environmentally Significant Behavior', *Journal of Social Issues*, 56/3, 407–24.

Stevenson, A. (2011) 'Coalition's Carbon Cuts Contain Get-Out Clause', http://www.politics.co.uk (home page), date accessed 13 May 2011.

Strøm, K. (1990) 'A Behavioural Theory of Competitive Political Parties', *American Journal of Political Science*, 34/2, 565–98.

Sutton, P. W. (2004) *Nature, Environment and Society* (Basingstoke: Palgrave Macmillan).

Sutton, P. W. (2007) *The Environment: A Sociological Introduction* (Cambridge: Polity Press).

Sunderlin, W. D. (2003) *Ideology, Social Theory and the Environment* (Lanham, MD: Rowman & Littlefield).

Tarrow, S. (1994) *Power in Movement: Social Movements, Collective Action, Collective Action and Politics* (Cambridge and New York: Cambridge University Press).

Tarrow, S. (2005) *The New Transnational Activism* (New York: Cambridge University Press).

Taylor, M., S. Rogers and P. Lewis (2011) 'Young, Poor and Unemployed: The True Face of England's Rioters', *The Guardian*, 19 August, 4–5.

TEE (2009) 'Roundtable Discussion', http://library.tee.gr/digital/m2482/m2482_contents.htm, date accessed 12 April 2010.

TELCO (2011) 'TELCO Citizens', http://www.citizensuk.org/chapters/telco/, date accessed 14 January 2012.

Telloglou, T. (2004) *The City of the Games* (Athens: Estia).

Thornburgh, N. (2011) 'London's Long Burn', *Time*, 22 August, 16–19.

Toohey, K. and A. J. Veal (2007) *The Olympic Games: A Social Science Perspective*, 2nd edn (Cambridge, MA: CABI Pub).

Totsikas, P. (2004) *The Other Facet of the Olympiad, Athens 2004* (Athens: Kappa, Psi, Mi).

Toyne, P. (2009) 'London 2011—Winning the Olympic "Green" Medal' in G. Poynter and I. MacRury (eds) *Olympic Cities: 2012 and the Remaking of London* (Farnham: Ashgate), pp. 231–42.

Tranter, B. (2011) 'Political Divisions Over Climate Change and Environmental Issues in Australia', *Environmental Politics*, 20/2, 78–96.

Tsiantar, D. (2010) 'Greek Tragedy: Athens' Financial Woes' *Time*, http://www.time.com/time/magazine/article/0,9171,1959059,00.html, date accessed 26 March 2010.

TWA [Tibetan Women's Association] (n.d.) 'China uses Tibetan Antelope as their Olympic Mascot' http://www.tibetanwomen.org/campaigns/beijing_olympics/, date accessed 16 September 2009.

UKELA (2008) *National Designated Sites*, http://www.environmentlaw.org.uk/rte.asp?id=204, date accessed 12 March 2009.

UNEP (2007) *Beijing 2008 Olympic Games—An Environmental Overview* (Nairobi: United Nations Environmental Programme).

UNEP (2009) *Independent Environmental Assessment: Beijing 2008 Olympic Games* (Nairobi: United Nations Environmental Programme).

UNSD (2009) *Greenhouse Gas Emissions: CO2 Emissions in 2006*, http://unstats.un.org/unsd/ENVIRONMENT/air_co2_emissions.htm, date accessed 15 January 2010.

UNSD (2010) 'Carbon Dioxide Emissions (CO2), Thousand Metric tons of CO2 (CDIAC)', http://unstats.un.org/unsd/mdg/SeriesDetail.aspx?srid=749, date accessed 15 March 2011.

Van der Heijden, H. A. (1999) 'Environmental Movements, Ecological Modernisation and Political Opportunity Structures' in C. Rootes (ed.) *Environmental Movements: Local, National and Global* (London: Frank Cass), pp. 199–221.

Van der Heijden, H. A., R. Koopmans and M. G. Giugni (1992) 'The West European Environmental Movement', *Research in Social Movements, Conflict and Change*, Supplement 2, 1–40.

Van Liere, K. D. and R. E. Dunlap (1980) 'The Social Bases of Environmental Concern: A Review of Hypotheses, Explanations and Empirical Evidence', *The Public Opinion Quarterly*, 44/2, 181–97.

Van Liere, K. D. and R. E. Dunlap (1981) 'Environmental Concern: Does it Make a Difference how it is Measured?', *Environment and Behavior*, 13/6, 651–76.

Veblen, T. (2005) *Conspicuous Consumption* (London: Penguin).

Vidal, J. (2010) 'UK Government Confirms Forest Sell-Off Plans', *The Guardian*, http://www.guardian.co.uk (home page), date accessed 30 October 2010.

Vigor, A., M. Mean and Ch. Tims (2004) 'Introduction' in A. Vigor, M. Mean and Ch. Tims (eds), *After the Gold Rush. A sustainable Olympics for London* (London: ippr and Demos), pp. 1–30.

Wachman, R. (2010) 'Is Greece the Eurozone's Canary in the Coalmine', *The Guardian*, http://www.guardian.co.uk (home page), date accessed 15 May 2010.

Walsh, B. (2009) 'Why Global Warming May be Fuelling Australia's Fires', *Time*, 9 February.

Watts, J. (2011a) 'Amid a Boom Built on Dirty Industry, China Plots Course for Green Growth', *The Guardian*, 5 February, 29.

Watts, J. (2011b) 'China Aims at Slower GDP Growth to Avoid Sacrificing Environment', *The Guardian*, 1 March, 23.

WCED (1987) *Our Common Future* (Geneva: World Commission on Environment and Development).

Webb, F. (2011) 'Environmental Protection UK Forced to Shut', *The Guardian*, http://www.guardian.co.uk (home page), date accessed 30 November 2011.

Weber, Max (1978) *Economy and Society: An Outline of Interpretative Sociology*, G. Roth and C. Wittich eds (New York: Oxford University Press).

Weidner, H. (2002) 'Capacity Building for Ecological Modernization: Lessons from Cross–National Research', *American Behavioral Scientist*, 45/9, 1340–68.

Weidner, H. and M. Jänicke, eds (2002) *Capacity Building in National Environmental Policy: A Comparative Study of 17 Countries* (New York/Berlin: Springer-Verlag).

Weston, J. (2002) 'From Poole to Fulham: A Changing Culture in UK Environmental Impact Assessment Decision Making?' *Journal of Environmental Planning and Management*, 45/3, 425–43.

Wiesenthal, H. (1993) 'Green Rationality' in J. Ferris (ed.), *Realism in Green Politics*, (Manchester: Manchester University Press), pp. 28–73.

Wilson, L. and M. Chambers (2011) 'Julia Gillard Seals Carbon Tax but Greens Want More', *The Australian*, http://www.theaustralian.com.au (home page), date accessed 25 November 2011.

Worden, M. (2008) 'Overview: China's Race for Reform' in M. Worden (ed.) *China's Great Leap. The Beijing Games and Olympian Human Rights Challenges* (New York, London, Melbourne, Toronto: Seven Stories Press), pp. 25–38.

WWF-Greece (2004) *Environmental Assessment of the Athens 2004 Olympic Games* (Athens: WWF).

Wynne, B. (1996) 'May the Sheep Safely Graze? A Reflexive View of the Expert-Lay Knowledge Divide' in S. Lash, B. Szerszynski and B. Wynne (eds), *Risk, Environment and Modernity: Towards a New Ecology* (London: Sage), pp. 44–83.

Xie, L. (2009) *Environmental Activism in China* (London and New York: Routledge).

Xu, G. (2008) *Olympic Dreams: China and Sports, 1895–2008* (Cambridge, MA: Harvard University Press).

Xu, X. (2006) 'Modernizing China in the Olympic Spotlight: China's National Identity and the 2008 Beijing Olympiad' in J. Horne and W. Manzenreiter (eds) *Sports Mega-Events: Social Scientific Analyses of a Global Phenomenon* (Oxford: Blackwell), pp. 90–107.

York, R. and E. A. Rosa (2003) 'Key Challenges to Ecological Modernization Theory', *Organization and the Environment*, 16/3, 273–88.

Young, D. (2005) 'From Olympia 776 BC to Athens 2004: The Origins and Authenticity of the Modern Olympic Games' in K. Young and K. B. Wamsley (eds), *Global Olympics: Historical and Sociological Studies of the Modern Games* (London: Elsevier), pp. 3–18.

Zhang, A. (2008) *The Environment, Beijing and the 2008 Olympic Games* (Beijing: Greenpeace).

Zhang, Y. (2010) 'The Future of the Past: On Wang Hui's Rise of Modern Chinese Thought', *New Left Review*, 62, 47–83.

Index

CPSIA information can be obtained at www.ICGtesting.com
Printed in the USA
BVOW04*2223071013

332999BV00003B/11/P